A Cost Estimating Framework for U.S. Marine Corps Joint Cyber Weapons

BRADLEY WILSON, THOMAS GOUGHNOUR, MEGAN MCKERNAN,
ANDREW KARODE, DEVIN TIERNEY, MARK V. ARENA,
MICHAEL J. D. VERMEER, HANSELL PEREZ, ALEXIS LEVEDAHL

Prepared for the United States Marine Corps Systems Command
Approved for public release; distribution unlimited

NATIONAL DEFENSE RESEARCH INSTITUTE

For more information on this publication, visit **www.rand.org/t/RRA1124-1**.

About RAND

The RAND Corporation is a research organization that develops solutions to public policy challenges to help make communities throughout the world safer and more secure, healthier and more prosperous. RAND is nonprofit, nonpartisan, and committed to the public interest. To learn more about RAND, visit www.rand.org.

Research Integrity

Our mission to help improve policy and decisionmaking through research and analysis is enabled through our core values of quality and objectivity and our unwavering commitment to the highest level of integrity and ethical behavior. To help ensure our research and analysis are rigorous, objective, and nonpartisan, we subject our research publications to a robust and exacting quality-assurance process; avoid both the appearance and reality of financial and other conflicts of interest through staff training, project screening, and a policy of mandatory disclosure; and pursue transparency in our research engagements through our commitment to the open publication of our research findings and recommendations, disclosure of the source of funding of published research, and policies to ensure intellectual independence. For more information, visit www.rand.org/about/principles.

RAND's publications do not necessarily reflect the opinions of its research clients and sponsors.

Published by the RAND Corporation, Santa Monica, Calif.
© 2023 RAND Corporation
RAND® is a registered trademark.

Library of Congress Control Number: 2023901826
ISBN: 978-1-9774-1020-7

Cover composite design: Carol Ponce; adapted from images by Nexusby/Adobe Stock, Ilya/Adobe Stock, Amgun/Getty Images.

About This Report

In support of U.S. Marine Corps Systems Command, RAND Corporation researchers assessed the Marine Corps offensive cyber operations acquisition life cycle and identified ways to improve the transparency of decisionmaking related to cyber weapon acquisition. The resulting recommendations focus on improving understanding of the tradespace of cost, schedule, operational capability, and risk to help the new Joint Cyber Weapons (JCW) program support and sustain the Marine Corps' cyber needs.

The research brought together data on operational capability, schedule, and risk into a cost-estimating framework with five classes of inputs and three types of outputs. The framework captured demand for exploits, various types of cyber weapons, the target environment (e.g., desktop or mobile systems), vulnerability decay rates, adversary defense capabilities, weapon cost, and how acquisitions are phased in and out of service over time. The analysis also involved quantifying vulnerability life spans and simulating the cost-estimating framework to better capture uncertainty.

This report presents the findings and recommendations, along with a RAND-developed cost-estimating framework that considered cost, schedule, operational capability, and uncertainty for the JCW acquisitions.

The research reported here was completed in June 2022 and underwent security review with the sponsor and the Defense Office of Prepublication and Security Review before public release.

RAND National Security Research Division

This research was sponsored by U.S. Marine Corps Systems Command and conducted within the Navy and Marine Forces Center of the RAND National Security Research Division (NSRD), which operates the National Defense Research Institute (NDRI), a federally funded research and development center sponsored by the Office of the Secretary of Defense, the Joint Staff, the Unified Combatant Commands, the Navy, the Marine Corps, the defense agencies, and the defense intelligence enterprise.

For more information on the RAND Navy and Marine Forces Center, see www.rand.org/nsrd/nmf or contact the director (contact information is provided on the webpage).

Acknowledgments

We thank David Pasquill, Program Management Office, Marine Corps Cyberspace Operations, for his guidance and insight during the development of the cost-estimating framework described in this report.

We would also like to thank Tim Conley and Trey Herr for their constructive and thorough reviews of this analysis. Finally, Paul DeLuca and Brendan Toland, Director and Associate Director, respectively, in the Navy and Marine Forces Center, provided valuable guidance, as well as insightful comments on the research.

Summary

Motivation

U.S. Marine Corps Systems Command asked the RAND Corporation to assess the Marine Corps offensive cyber operations acquisition life cycle and identify ways to improve the transparency of decisionmaking related to cyber weapon acquisition. The goal was to enhance understanding of the tradespace of cost, schedule, operational capability, and risk to help ensure that the new Joint Cyber Weapons (JCW) program can best support and sustain the Marine Corps' cyber needs.

Approach

To increase transparency and inform decisions about JCW's pending acquisition, we brought together data on operational capability, scheduling, and risk to develop a JCW life-cycle cost-estimating framework. This framework should help program leadership understand the potential costs and provide additional guidance on budgeting considerations. It incorporates five classes of inputs and has three types of outputs, as summarized in Figure S.1.

We considered the demand for cyber weapons from the operational user, as well as the type of cyber weapon (e.g., exploit, implant, payload), the weapon's target environment (e.g., desktop or mobile systems), vulnerability decay rate, the adversary's defense capabilities, weapon cost, and how various acquisitions are phased in and out of service over time. We computed the production of cyber weapons, their costs, and how uncertainties are distributed over a specified period (e.g., obligation authority for the Future Years Defense Program). The framework was implemented and delivered to the sponsor in Microsoft Excel so they could best make use of it.

In developing our framework, we uncovered numerous potential uncertainties related to the life span of vulnerabilities, the development demand signal, and exploit acquisition approaches, as well as challenges with data availability. For these reasons, we attempted to quantify vulnerability life spans and build a simulation of the cost-estimating framework in R to provide improved transparency and a more granular view of uncertainty in areas that are important to the JCW program.

Findings

In addition to uncovering numerous potential uncertainties in estimating the cost of cyber weapons, we assessed the life spans of 133 historic vulnerabilities using open-source information and found that the mean life span can be quite short for mobile and desktop vulner-

FIGURE S.1

Scope of the JCW Life-Cycle Cost Estimating Framework

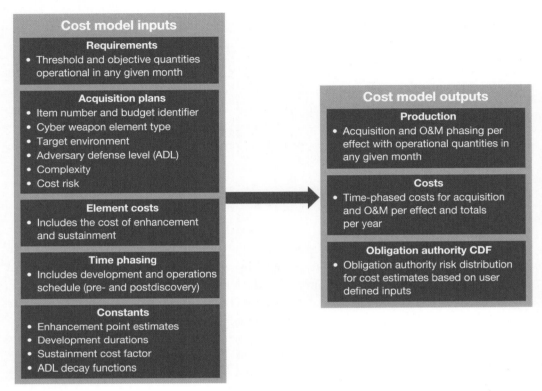

NOTE: CDF = cumulative distribution function; O&M = operations and maintenance.

abilities (three to five months, respectively) in situations in which potential adversaries have a high defense level (i.e., an ability to rapidly identify and patch a vulnerability).

Based on the available data and assumptions about operational demand, we found that there is significant uncertainty in the potential cost of the JCW program—a five-year total cost between $90 million and $290 million.

It is important to note that the operational demand level is notional, as is the cost of procuring and operationalizing an exploit, and costs could change as more data become available. There are also other parameters that could be explored, such as the complexity of exploits.

The cost-estimating framework presented in this report represents a foundation for what we hope will be incremental improvements as understanding of the challenges improves and as additional historical data become available.

Recommendations and Next Steps

Our findings point to the following recommendations for JCW program leaders:

- Consider the significant uncertainty of the life span of vulnerabilities during program planning and budgeting.
- Collect historical data (and plan to collect future data) on the cost and schedule of procuring and operationalizing exploits.

Over time, as more acquisition and execution data become available, the Marine Corps can update the model and inputs to better predict future costs. We have identified seven areas to enhance the model: (1) incorporate historical data on cyber weapon element cost, (2) increase the granularity of cyber weapon element estimates, (3) increase the granularity of cyber weapon types, (4) characterize the acquisition type for the cyber weapon, (5) incorporate additional cost and schedule drivers, (6) further develop and parameterize the adversary defense level decay functions for exploits, and (7) convert the exploratory simulation into an optimization model.

Contents

Figures and Tables

Figures

Tables

CHAPTER ONE

Introduction

Because cyberspace operations have become increasingly common and complex, the U.S. Department of Defense (DoD) needs to develop a better understanding of what it needs to acquire and the cost, schedule, and performance of the acquisitions. RAND Corporation was asked to assess the Marine Corps cyber operations acquisition life cycle and identify ways to improve the transparency of connections between Marine Corps Cyber Operations Program Management Office (PMM-173) capabilities and strategic guidance. RAND was further asked to provide recommendations to enhance cyber weapon acquisition decisionmaking by improving understanding of the tradespace of cost, schedule, operational capability, and risk to help ensure that the new Joint Cyber Weapons (JCW) program can best support and sustain the U.S. Marine Corps' cyber needs. This report presents a framework to estimate the cost of cyber weapons within the context of schedule, capability, and risk. We also present an exploratory simulation capability designed to improve transparency by quantifying uncertainty in important areas.

Joint Cyber Weapons Program

This work was executed in support of the new Marine Corps JCW program. The program is executed by U.S. Marine Corps Systems Command in support of the Marine Corps Forces Cyberspace Command. JCW provides the command with offensive cyber capabilities for use in their cyber operations.

The nature of offensive cyber operations means that the capabilities acquired by JCW are nontraditional in a few important ways:

- They can depend on markets where dollar values for capabilities can be difficult to define.
- They need to be developed very quickly (i.e., in six months or less).
- They need to be fully capable by the time they are deployed using agile software development methods.
- Their developers do not need to consider user or data migration needs.
- They usually do not have a long life span relative to a typical acquisition, making sustainment tails small to nonexistent.
- Opportunities to reuse elements of previous projects are limited.

The cost model framework described in this report was designed to be responsive to some of these aspects, specifically the rapid development period and relatively short life span of the cyber weapon.

Software Acquisition Pathway

The JCW program uses the software acquisition pathway (SWP) within the Adaptive Acquisition Framework. JCW was officially designated on the SWP in October 2021 (Pasagian, 2021). The SWP was added to DoD Instruction 5000.02, *Operation of the Adaptive Acquisition Framework*, in January 2020, after being prescribed in Section 800 of the National Defense Authorization Act for fiscal year (FY) 2020 (Pub. L. 116-92, 2019). The pathway was further articulated in DoD Instruction 5000.87, *Operation of the Software Acquisition Pathway*, in October 2020. The pathway provides JCW with mechanisms to deliver software rapidly by flexibly managing requirements, along with an ability to customize software engineering life-cycle practices, processes, and associated artifacts.

The SWP is new and, given the uniqueness of the JCW program relative to traditional software acquisitions, might face some implementation challenges. For example, SWP policy is evolving, and increasing congressional interest could prompt greater scrutiny and a larger number of data calls. However, perhaps the most challenging aspect is that cyber weapons need to be fully operational by deployment. The concept of a minimum viable product that users can then implement operationally is likely not satisfactory. This increases the importance of the test and evaluation phase of development, which will challenge the JCW program's ability to rapidly deliver effects.

Cyber Weapon Terminology

Before discussing the details of the cost-estimating framework, it is important to define some relevant terms. Cyber weapons are developed to attack adversaries' systems through a multistep process that exploits underlying vulnerabilities in the target system. Several organizations have tried to define the multistep process of sophisticated cyberattacks, often to assist others in providing comprehensive network defense against cyber weapons. Lockheed Martin described a seven-step "cyber kill chain" that began with reconnaissance of target systems and ended with actions against objectives (Lockheed Martin, 2015). MITRE has created an attack framework, which includes matrixes of techniques that are applicable to different groups of platforms (MITRE, undated c). The framework for enterprise platforms comprises 14 groups of techniques associated with steps in a cyberattack, beginning with reconnaissance and ending with impact (MITRE, undated c).

The focus of our cost-estimating framework lies within these holistic process descriptions and encompasses three general categories of tools that ultimately make up a cyber weapon: exploits, implants, and payloads. Throughout this report, we refer to these categories as the *elements* of a cyber weapon; when we discuss *element types*, we are referring to exploits, implants, or payloads. We describe each element type in more detail later in this chapter.

Beyond our focus on cyber weapon elements, the cyber kill chain and the MITRE framework both begin with the reconnaissance step. This is because the elements of a cyber weapon have utility only after one or more vulnerabilities in the target system have been identified. The recent widespread cyberattacks on Microsoft Exchange Servers illustrate how a cyber weapon can be used to exploit vulnerabilities and achieve an attacker's desired effects. In March 2021, Microsoft announced that hackers were using vulnerabilities in Microsoft Exchange Servers to access email accounts and install malware on affected systems (Microsoft, 2021a). The following subsections will provide definitions and context for the vulnerabilities and each of the elements of a cyber weapon, and each concludes with an illustrative example based on the attack on Microsoft Exchange Servers.

Vulnerabilities

When discussing cyber weapons, we need a shared understanding of the basic concept of a vulnerability. The Common Vulnerabilities and Exposures (CVE) program defines *vulnerability* as follows:

> A flaw in a software, firmware, hardware, or service component resulting from a weakness that can be exploited, causing a negative impact to the confidentiality, integrity, or availability of an impacted component or components. (MITRE, undated b)

The cyber weapons discussed in this report are tools developed to exploit such vulnerabilities to achieve a desired effect. Where these vulnerabilities arise from flaws in code, they can often be fixed by replacing the code with a version that does not include the flaw—a process known as *patching*. When a vulnerability is discovered, it behooves users to install available patches before the vulnerability can be exploited. Many software components contain undiscovered vulnerabilities, however, and a patch might not be immediately available when a vulnerability is first discovered (or first exploited by an attacker). These are called *zero-day vulnerabilities*. The term *zero-day* refers to the number of days a software vendor has known about the vulnerability (Libicki, Ablon, and Webb, 2015). This concept can be expanded to a more generic term, *n-day vulnerabilities*, where *n* is the number of days a vendor has known about the vulnerability.

The significance of vulnerabilities varies considerably according to what effects they may enable an attacker to achieve and how challenging it is to exploit them. Many databases and taxonomies have been created to describe, categorize, and enumerate existing vulnerabilities, and the appendix presents a more detailed discussion.

In the case of the attacks on Microsoft Exchange Servers, at least four zero-day vulnerabilities were used in concert. The first vulnerability allowed attackers to forge authentication as an Exchange Server. The remaining three vulnerabilities could have provided the attackers with the ability to write files to any path on the server and execute arbitrary commands.

The latter three vulnerabilities were enabled only once the attacker had forged authentication by exploiting the first vulnerability (Microsoft, 2021a). Microsoft was able to quickly release patches for these vulnerabilities, but these only prevented the vulnerabilities from being further exploited and did not address situations in which an attacker had already established a persistent presence on a network (Microsoft, 2021b).

The next three subsections describe the cyber weapon elements of interest in our cost-estimating framework.

Exploits

If a *vulnerability* is defined as a flaw in a software, firmware, or service component resulting from a weakness that can be exploited, an *exploit* is the code that takes advantage of that flaw to achieve a negative impact on the confidentiality, integrity, or availability of the component (MITRE, undated b). Put more simply, an exploit is the program or block of code written to take advantage of a vulnerability and establish or expand unauthorized access or privileges in a target system.[1] Exploits, by definition, exist in only the context of the vulnerability they are designed for. An exploit cannot exist without a vulnerability, although, historically, most vulnerabilities do not have known exploits (Ablon and Bogart, 2017; Ballard, 2021). Attempts to characterize exploits therefore often overlap with efforts to track, describe, and categorize vulnerabilities and are often discussed only in the context of the defined vulnerabilities (e.g., when an attacker exploits a published common vulnerability or exposure).

In the context of cyber weapons, exploits are most closely associated with the concept of *access*. Exploits are the tools that establish an initial presence on a target system; expand access to other areas of a network; and provide the attacker with enhanced privileges to perform additional actions, such as file modification or arbitrary command execution (Ablon and Bogart, 2017). They are often described or categorized according to what they allow an attacker to do (e.g., remote code execution [RCE] or sandbox escape; see Zerodium, undated), but such a categorization would resemble an enumeration of the actions that any user could perform on the target operating system (OS). The appendix discusses some categories of exploits in more detail, but for the purposes of this report, it is most useful to consider exploits as simply the tools that confer the access and privileges necessary to install implants and deliver payloads. In our framework, which we introduce in Chapter Two, we do not consider exploits at a lower level of detail, although we might do so in future versions of our cost-estimating framework.

Cyber weapons can—and often do—use multiple exploits in a chain to gain an attacker the level of access and system privileges needed to achieve desired effects. Such a chain of

---

[1] The modifier *unauthorized* is a key word that distinguishes an exploit from other legitimate tools or techniques allowing the same effects. An alternative definition for *exploit* comes from the Cybersecurity Glossary of the National Initiative for Cybersecurity Careers and Studies: "A technique to breach the security of a network or information system *in violation of security policy*" (emphasis ours) (National Initiative for Cybersecurity Careers and Studies, 2022).

exploits was used in the case of the Microsoft Exchange Server attacks. Attackers exploited a vulnerability that allowed them to authenticate as an Exchange Server, providing initial access. This initial forged authentication enabled them to execute three other exploits that conferred the privileges needed to write files to any path and execute arbitrary commands. This chain of exploits allowed the attackers to create implants, in the form of web shells (described in the next section), which facilitated further actions even after the exploited vulnerabilities were patched (Microsoft, 2021b).

Implants

An *implant* is "a program that solidifies and maintains access initially provided by an exploit (i.e., achieves persistence) and delivers some effect to the system" (Ablon and Bogart, 2017). In the context of cyber weapons, implants are therefore most closely associated with the concept of *persistence*. If an exploit were analogized as an invasion, an implant would be the equivalent of a beachhead: the established area that allows the attacker to maintain access without needing to reinvade with each action and the point from which additional reconnaissance can be performed and further attacks can be launched. These further attacks could be either additional exploits that expand access and privileges or the delivery of the payload(s) that achieve the intended effects of the cyber weapon. Although the functions of exploits and implants are occasionally very similar, the most useful difference between the two elements pertains to reliability. Exploits are dependent on vulnerabilities that can be ephemeral, and the effectiveness of exploits can vary considerably, depending on the characteristics of the target. The reliability of exploits is, therefore, often uncertain. An implant, in contrast, is required and designed to be persistent and reliable to ensure that post-exploitation activities can continue in the target environment, including the eventual delivery of the payload (Ablon and Bogart, 2017).

Implants are often able to ensure access and control even after the vulnerabilities that allowed the initial exploits are patched. This was the case in the Microsoft Exchange Server attacks. The initial chain of exploits the attackers used enabled them to create implants in the form of web shells on infected servers. These web shells created back doors that allowed the attackers to retain the access and privileges gained from the exploit chain without needing to perform the exploits again. Microsoft released patches that addressed the vulnerabilities, but these patches did not clean these web shells from infected servers. Attackers were able to maintain access to patched systems and, eventually, use these web shells to deploy a variety of payloads (Microsoft, 2021b).

Payloads

The *payload* is the element of a cyber weapon that achieves the weapon's intended effect (Bellovin, Landau, and Lin, 2017). Although all elements of cyber weapons are code written to achieve some specific purpose and although the line between the payload and the other elements (especially exploits) can be blurry, payloads are most clearly distinguished from the other elements by the achievement of a specific objective, usually the terminal objective. In some cases, there might be no difference between the exploit and the payload when a single exploit or a chain immediately achieves the intended effect. Payloads may also be executed with a variety of triggers. Some are self-replicating, automated, and untargeted and are triggered as soon as they are initiated, but such indiscriminate weapons are prohibited as instruments of war (Office of the General Counsel, 2016; Bellovin, Landau, and Lin, 2017). Targeted payloads could still be triggered without active intervention (also known as *hands on keyboard*), such as when the trigger is scheduled or automatically executed when certain target system configurations occur. This was the case with Stuxnet, which executed its payload only after detecting a certain plant configuration (Bellovin, Landau, and Lin, 2017). Manual payloads that are triggered by hands on a keyboard allow the most controlled targeting of effects and are often highly dependent on a reliable implant for success. Payloads can be designed to achieve a wide variety of effects, including the following:

- denial of service
- data monitoring or exfiltration
- file manipulation, corruption, encryption, or destruction
- repurposing system processes to perform other actions for an attacker (e.g., setting up botnets or acting as command and control for further actions)
- disabling an OS
- destruction of a cyber-physical system (potentially including collateral damage of nearby property or personnel).

In the case of the Microsoft Exchange Server attacks, many different attackers deployed varying payloads to achieve their ends. The web shell implants allowed attackers to perform hands-on-keyboard actions to trigger payloads. Many attackers used web shells to exfiltrate data, then encrypt system files to extort a ransom payment. Exfiltrated data often included credentials stored on a server that could be used to facilitate later attacks. Other attackers deployed payloads that repurposed target systems to perform cryptocurrency mining operations, also known as *cryptojacking*. Repurposed systems were also often used for command and control of further payload dissemination (Microsoft, 2021b).

Now that we have established some of the key terminology, we describe considerations for estimating the costs for the procurement or development of these cyber weapon elements.

Challenges in Estimating the Cost of Cyber Weapons

In most applications, estimating the cost of a program or project is straightforward, given known requirements and data. In practice, even when cost estimation seems like a trivial task, it can be challenging to determine resource needs and technical requirements and to collect the requisite data to effectively estimate the cost of an acquisition. These challenges are exacerbated in the case of estimating the cost of cyber weapons. In this section, we discuss some of the reasons.

Uncertain Life Spans of Vulnerabilities

Perhaps the most challenging aspect of developing a cost estimate for the JCW program is the uncertainty of the duration of a given vulnerability. Vulnerabilities can be discovered at any time, rendering moot an investment in a cyber weapon that exploits the vulnerability (as users patch their affected software over time). Without a known life span for what is being acquired, the demand signal for the portfolio of weapons the program may need is uncertain.

A vulnerability can be identified and defended through a security patch quite quickly, diminishing the operational capability of a cyber weapon designed to target unpatched adversary systems. As adversaries discover vulnerabilities and defend against them, the utility of the cyber weapons that exploit the vulnerabilities will decay. Taken to the extreme, it is conceivable that—from the time a cyber weapon requirement is identified to when the weapon is developed and ready to be deployed—the weapon may no longer be useful at all. However, not all adversaries are equal, and different adversaries might patch at different rates. Costs are always incurred, but the realization of benefit often depends on the adversary. This risk should be considered in JCW cost-estimating and budgeting activities.

Uncertain Development Demand Signal

Given the uncertainty of vulnerability life spans, it is difficult to know the number of weapons that should be in development at any given time or the development demand signal. The operational goal for the number of weapons is likely more straightforward (e.g., the operators need to have four weapons available at any given time). How many additional weapons need to be available and in development? In our cost-estimating framework, we propose options to answer this question and manage uncertainty that vary by the type of vulnerability, the type of adversary, estimated decay rate, and cost.

Limited Data Availability

Data availability is not an uncommon challenge when developing a cost estimate. However, it is particularly challenging in the domain of cyber weapon cost estimation for at least a few reasons. First, the development of offensive cyber weapons is relatively new, and minimal historical cost and schedule data exist. Second, where cost and schedule data do exist, they are proposal or contract data, not execution data. Finally, even when data are available, it is difficult to persuade outside organizations to share them because of the sensitivity around cyber warfare and cyber weapon development.

The means of development could include purchasing exploits from a vendor in a highly specialized and opaque market, and such vendors may not wish to disclose their cost data. Although it might be easier to acquire and use data from cyber weapon research and development conducted completely in house, these data will not always be useful as an analogy for exploits purchased on the gray or white markets.[2] These challenges lead to other cost-estimating issues even when some cyber weapon cost data exist. The cost estimator must use caution in the absence of contextual data for a weapon. If the cost estimator has only cost data but does not know how the weapon was developed or its characteristics, the cost data may not be an appropriate analogy for future cyber weapon development and could be erroneously applied.

Uncertain Exploit Acquisition Approach

As noted, the process for identifying a vulnerability differs depending on whether a weapon will be purchased, a result of in-house development, or a mix of the two. The approach might not be known in advance of estimating the cost of cyber weapons, where a key cost element is operationalizing the exploit to take advantage of a vulnerability. Once the requirement for a particular cyber weapon is determined, some research will likely occur early in the development process to identify the best acquisition approach, whether procurement from a vendor or in-house development. Each approach may have different cost considerations. Unfortunately, cost estimates for budgets need to occur before the final acquisition approach is determined, resulting in a need to make estimating assumptions and increased uncertainty in the cost estimates. That said, the analytic approach described can help the JCW program better manage this uncertainty.

Organization of This Report

To address these challenges, we propose a cost model for the JCW program that incorporates estimation and budgeting in Chapter Two. We demonstrate an exploratory simulation and analysis approach that implements the cost model and expands on it to help the JCW program quantify areas of uncertainty in Chapter Three. We discuss future model enhancements and offer recommendations in Chapter Four. The appendix discusses some categories of exploits in more detail.

The cost model and analysis documented in this report can provide the JCW program with a clearer view of the connection between estimated costs and capability needs. Collectively, they provide a robust foundation for a data-driven approach to cost estimating that also accounts for scheduling, operational outcomes, and risk and uncertainty.

---

[2] A *gray market* is a place where zero-day exploits are sold "frequently in weaponized form to government agencies and other buyers seeking to deploy them for offensive purposes" (Stockton and Golabek-Goldman, 2013, p. 247).

CHAPTER TWO

Cost-Estimating Framework

This chapter discusses our cost-estimating framework, which consists of a model and tool used to estimate the costs and budgets for the types of cyber weapons discussed in Chapter One. We developed the tool in Microsoft Excel at the request of the sponsor for its familiarity and compatibility with prevailing frameworks and processes across DoD.

As with any cost-estimating framework or model, there are basic inputs and outputs. Figure 2.1 provides a high-level depiction of five inputs and three outputs of the cost model. The inputs include requirements, acquisition plans, element costs, time-phasing, and constants. The outputs include production, costs, and an obligation authority cumulative distribution function (CDF). We discuss each of these model sections in turn.

Inputs

Requirements

High-level requirements are the first user input. This portion of the model simply provides a free-form input of requirement counts for the three types of cyber weapon elements: exploits, implants, and payloads. Here, the user inputs the threshold and objective quantity, or operational demand, for each element type over time.[1] The model allows the user to enter monthly quantities. Although it is perhaps unusual to track acquisition requirements monthly, the acquisition and development tempo for cyber weapons is much faster than for traditional weapon systems. This is a function of the uncertain and potentially short life span of cyber weapons and of the fluid and fast-paced nature of cyber warfare. It is more common to plan for and track development costs on a quarterly or annual basis for traditional weapon systems, such as tanks, missiles, and aircraft, or even other types of information technology, such as enterprise resource management systems. A key difference is that these types of weapon sys-

---

[1] Operational demand is distinct from the development demand signal. *Operational demand* is the requirement or desired objective quantity of cyber weapon elements. The development demand signal is determined based on model outputs and takes into account the uncertainties described in Chapter One. The resulting development demand signal quantity will almost always be greater than the requirement goal, a result of decay or obsolescence.

FIGURE 2.1

Scope of the Life-Cycle Cost Estimating Framework

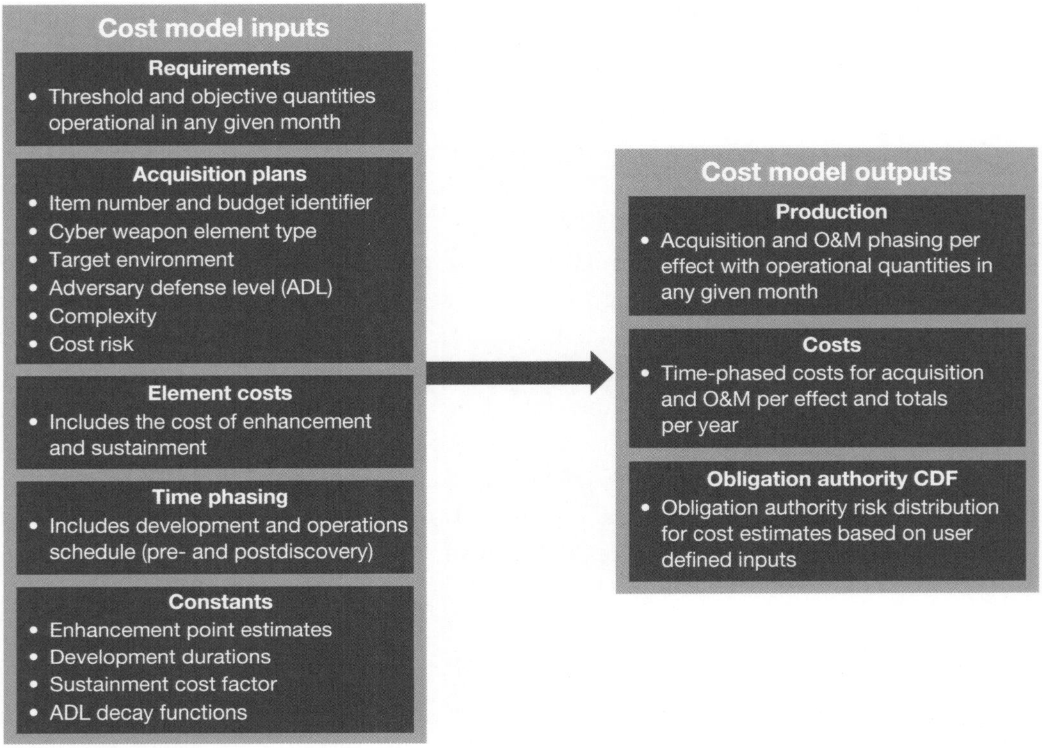

Cost model inputs

Requirements
- Threshold and objective quantities operational in any given month

Acquisition plans
- Item number and budget identifier
- Cyber weapon element type
- Target environment
- Adversary defense level (ADL)
- Complexity
- Cost risk

Element costs
- Includes the cost of enhancement and sustainment

Time phasing
- Includes development and operations schedule (pre- and postdiscovery)

Constants
- Enhancement point estimates
- Development durations
- Sustainment cost factor
- ADL decay functions

Cost model outputs

Production
- Acquisition and O&M phasing per effect with operational quantities in any given month

Costs
- Time-phased costs for acquisition and O&M per effect and totals per year

Obligation authority CDF
- Obligation authority risk distribution for cost estimates based on user defined inputs

NOTE: O&M = operations and maintenance.

tems and information technology are unlikely to be obsolete when they become operational, but this is a possibility for cyber weapons.

Threshold requirement counts represent the minimum number of cyber weapon elements required to meet mission objectives, and objective requirement counts represent the goal or desired number of cyber weapon elements. Figure 2.2 shows a notional example of inputs on the cost model's requirement input worksheet. In this example, the monthly threshold requirement for exploits is three throughout FYs 2024 and 2025.

It is important to note that these are cumulative counts, not incremental counts. Therefore, for exploits, the operational threshold is to maintain three exploits throughout FYs 2024 and 2025. The model is not incrementally adding three additional exploits every month. Also note that the requirement inputs worksheet merely shows exploit, implant, and payload counts. This worksheet provides no specific information about the characteristics of these elements. The cyber weapon element characteristics are captured on the acquisition plans worksheet.

FIGURE 2.2

Requirements Worksheet

Cyber Weapon Element Requirement Inputs	Requirement Type	FY 2024												FY 2025											
		Oct-23	Nov-23	Dec-23	Jan-24	Feb-24	Mar-24	Apr-24	May-24	Jun-24	Jul-24	Aug-24	Sep-24	Oct-24	Nov-24	Dec-24	Jan-25	Feb-25	Mar-25	Apr-25	May-25	Jun-25	Jul-25	Aug-25	Sep-25
Operational Exploit Requirement/ Goal	Threshold	3	3	3	3	3	3	3	3	3	3	3	3	3	3	3	3	3	3	3	3	3	3	3	3
Operational Exploit Requirement/ Goal	Objective	4	4	4	4	4	4	4	4	4	4	4	4	4	4	4	4	4	4	4	4	4	4	4	4
Operational Implant Requirement/ Goal	Threshold	3	3	3	3	3	3	3	3	3	3	3	3	3	3	3	3	3	3	3	3	3	3	3	3
Operational Implant Requirement/ Goal	Objective	4	4	4	4	4	4	4	4	4	4	4	4	4	4	4	4	4	4	4	4	4	4	4	4
Operational Payload Requirement/ Goal	Threshold	3	3	3	3	3	3	3	3	3	3	3	3	3	3	3	3	3	3	3	3	3	3	3	3
Operational Payload Requirement/ Goal	Objective	4	4	4	4	4	4	4	4	4	4	4	4	4	4	4	4	4	4	4	4	4	4	4	4

Acquisition Plans

The acquisition plans serve as a running list of discrete cyber weapon elements to be developed. That is, each row represents one element and is counted as one requirement, as defined in the requirements portion of the cost model. There are three acquisition plan sections on the acquisition plans worksheet: requirement characteristics, costs, and phasing (or schedule). Figure 2.3 shows the requirement characteristics. Inputs in the requirement characteristic inputs section help with data organization and inputs that affect cost and/or schedule estimates. We discuss the organizational inputs first.

Organization

Organizational inputs include the item number and budget identification (ID) number. The item number input is a number assigned to each cyber weapon element requirement. It is a running list of the cyber weapon elements in the cost model and aids in tracking the number of items in the model and in linking items as they appear on other worksheets in the model.

The second organizational input is a budget ID number to be assigned to each cyber weapon element line item. This may be useful if cyber weapon elements get bucketed under different budget IDs and if the model user wishes to view total costs for a specific budget ID.

The remaining requirement characteristic inputs include fields that are a set of drivers affecting either the cost, schedule, or both. These include the cyber weapon element, the target environment, adversary defense level (ADL), and cyber weapon element complexity, along with a field to capture the cost risk associated with each cyber weapon element.

Cyber Weapon Element

The first cost driver is the cyber weapon element. The model allows the user to choose among three elements: exploit, implant, and payload. The costs can vary dramatically, depending on which element is being developed. Implants are much more costly to develop than exploits and payloads. This is partly due to the reliability requirements of implants relative to exploits and payloads. Implant reliability is critical because of the requirement to maintain access to

FIGURE 2.3

Requirement Characteristic Inputs

Item #1	Cyber weapon element	Target environment	Adversary defense level	Budget ID	Complexity	Cost risk

• 1–n	• Exploit
	• Implant
	• Payload
		• Desktop or server
		  – Windows
		  – macOS
		  – Linux
		  – Any OS
		  – Router
		• Mobile
		  – iOS
		  – Android
		  – Any OS
		  – Server
		  – Router
		• Not applicable to any OS
			• Low
			• High
				• Alphanumeric budget code	• Low
					• Medium
					• High
						• None
						• Low
						• Medium
						• High
						• Very high

an adversary target environment and persist in delivering some effect. Exploits thus provide initial access to a target, while implants maintain that access. It is more costly, on average, to develop implants with the level of reliability required to maintain access (Ablon and Bogart, 2017).

Target Environment

The second cost driver is the target environment for each element. We drew from the Zerodium organization scheme, which distinguishes whether the target environment is a desktop or server or is a mobile device (Zerodium, undated).[2] Within these two broad categories, the target environment is broken down into the type of OS or the hardware (e.g., servers, routers) targeted. For desktops, targeted systems include Windows, macOS, and Linux. The model also allows for desktop targets, which may be OS-agnostic or not applicable to any OS. For mobile devices, targeted systems include Android or iOS, or, as with desktops, the user can choose an OS-agnostic option. Other types of hardware may be targeted as well. The model allows the user to select both desktop and mobile servers and routers as targets.[3]

Adversary Defense Level

The ADL for a given cyber weapon element is the third driver in the model and affects the schedule more directly. The current model framework allows the user to assume a low or high ADL. The cost driver is intended to capture the adversary's level of sophistication in defending against cyber weapons. For instance, an adversary with a high defense level would likely

---

[2] The Zerodium framework expands to lower levels, distinguishing between different types of web browsers, web servers, email servers, clients or files, and web apps with potentially exploitable vulnerabilities. This first iteration of the model limits the vulnerability distinctions to desktop or server and mobile OSs and hardware, but future model iterations could expand the target environment options to the lower levels identified in vulnerability frameworks, such as Zerodium.

[3] Future versions of the model will likely consider infrastructure, such as routers, separately.

be a nation-state or actor with considerable knowledge of cyber weapons and significant resources allocated to defending against cyberattacks. They are likely to identify and patch their systems more quickly, increasing the rate of decay for cyber weapons. On the other hand, an adversary with a low defense level might be a small terrorist organization without the knowledge and resources to ensure that n-day vulnerabilities are adequately patched, decreasing the rate of decay and, thus, potentially extending the life of a cyber weapon. In the model, we implemented and adjusted the slope of a decay function that models the useful life of a given cyber weapon element. Therefore, the decay rate of the element's usefulness will be slower (i.e., it will have less-steep curve) when the user assumes a low ADL for the element, extending the weapon's operational life. The implementation of the decay functions is discussed in the "Constants" section later in this chapter.

Complexity

A fourth cost driver is intended to capture the complexity of the element. The current framework rates complexity as low, medium, or high. Although most public information on exploit characteristics that affect prices pertains to the purchase of exploits on the gray market, it appears to be a reasonable assumption that characteristics that drive up market prices are correlated with the level of research and development necessary to exploit a vulnerability.

Ruef, 2016, proposes six characteristics that drive the cost of an exploit:

- *Target popularity.* Popular targets are likely better defended, therefore requiring more-complex exploits and placing a premium on procuring or developing exploits for popular targets. This is supported by Zerodium's published bounties from January 2019 (Zerodium, undated). The payout for a Chrome exploit was up to $500,000, and Safari, Edge, and Firefox had a ceiling of $100,000. This roughly correlates with the popularity of these browsers at the time, when the estimated desktop browser market share was 71 percent for Chrome, 10 percent for Firefox, 5 percent for Internet Explorer, 5 percent for Safari, and 4 percent for Edge (StatCounter, undated).
- *Exploit exclusivity.* If the exploit is sold exclusively to a single buyer, the price will be significantly higher than if the exploit is sold to multiple buyers, shortening the useful life of the exploit.
- *Exploit quality.* High-quality exploits may decrease detectability and be implemented in an elegant way that is easier for a user to understand.
- *Exploit reliability.* This is arguably part of exploit quality. Errors in complex weapons can cause targets to crash, alerting adversaries to attacks and giving them the opportunity to defend against the weapon.
- *Exploit penetration force.* The more layers of security an exploit must penetrate, the more sophisticated and extensive its development needs to be. Multiple exploits are often chained together to establish an implant, so exploits with a high penetration force would be less dependent on other exploits.

- *Payload options.* The effect enabled by an exploit is a factor in exploit complexity and, therefore, cost. For example, an exploit that only allows a denial-of-service attack will be less complex and costly than one that grants an attacker extensive write access in a database environment.

To date, we have not implemented the complexity factor as a function of these or any other set of potential dimensions. Our model is an abstract representation of the fact that not all exploits are created equal.

Cost Risk

Finally, the cost model allows the user to capture the expected cost risk of cyber weapon elements. The cost risk could be attributable to a multitude of uncertainties:

- Volatile labor rates and availability of labor expertise, particularly in the development of cyber weapons, can make it challenging to find qualified developers and relevant experts with the required security clearances.
- The fidelity of the requirements definition in the context of rapidly changing requirements. The fluid nature of cyber warfare can introduce uncertainty throughout a cyber weapon's life cycle.
- A cyber weapon could become unusable in a short time frame, and the time frame for cyber weapon decay is often uncertain.

The model has two options for characterizing the cost risk of each element. The first option is useful when specific historical data are not available to adequately quantify the expected cost risk of a given effect. This method leverages the *Joint Agency Cost Schedule Risk and Uncertainty Handbook* to quantify expected cost risk subjectively when historical data are not available (Naval Center for Cost Analysis, 2014). In the model, the user can select from five cost risk levels (see Figure 2.3), with options ranging from "none" (i.e., no cost risk) to "very high" (i.e., exceptional cost risk). Associated with each of these subjective options is a sigma value that represents the standard deviation of the uncertainty distribution in log space for the point estimate.

An alternative to using the cost risk field to capture uncertainty in the cost of the effect is to use the optional field labeled "risk override." This allows the user to input their own sigma value if they have the historical data to analyze and use as a more specific and relevant measure of cost risk for the given effect. If a value is entered into the risk override field, as suggested, this sigma value will override the option chosen in the cost risk field and will be used in the subsequent cost inputs field, as described next.

Element Costs

Following the requirement characterization input fields is the cyber weapon element cost section. The fields include each line-item enhancement cost (i.e., the development of the actual effect) and the monthly sustainment cost. The costs for both enhancement and sustainment include a point estimate, a mean estimate, and the standard deviation of the estimate. The enhancement point estimate draws from a lookup table on the constants tab, which captures the average point estimate cost for each element type, target environment, and complexity combination (e.g., an exploit for a high-complexity Microsoft Windows desktop target will have its own discrete point estimate cost).[4] The point estimate can be thought of as the most likely cost of the element.

The next field for each line-item element introduces the use of the cost risk input and captures the mean cost estimate for enhancement using a log-normal distribution and sigma value, which is determined by the cost risk or risk override user input. The log-normal mean equation used in the model is

$$lognormal\ mean\ estimate\ =\ point\ estimate * e^{\mu+\sigma^2/2},$$

where μ is 0 and σ is the standard deviation (or sigma)—both in log-space. In addition to the log-normal mean, the log-normal standard deviation is calculated using the following equation:

$$lognormal\ std\ deviation = \sqrt{e^{2\mu+\sigma^2}(e^{\sigma^2} - 1)*(point\ estimate)^2}.$$

Again, μ is 0 and σ is the standard deviation (or sigma)—both in log-space.

To describe the mechanics of the point estimate, mean estimate, and standard deviation calculations more clearly, we will go through a notional example. The point estimates for the various cyber weapon element characteristics are found on the constants worksheet. If, for instance, a user adds a new item to the acquisition plan inputs worksheet that is an exploit intended to target a Windows desktop and is of medium complexity, the user will choose these requirements characteristics and a point estimate for the development of the item will automatically be selected from a look-up table of point estimates which capture the point estimates for all characteristic combinations. In this case, an average Zerodium bounty price is drawn for $170,303 (base year 2021 dollars). This represents the point estimate or, equivalently, the most likely estimate. The mean estimate, unlike the point estimate, accounts for risk, and the expected amount of risk is determined by the user-defined sigma, as described earlier. In our example, suppose the user inputs a "very high" cost risk for the Windows desktop exploit we are estimating. A "very high" risk is associated with a sigma value of 0.45 in log-space. The model calculates the lognormal mean of the estimate with the mean estimate equation above.

$$lognormal\ mean\ estimate\ =\ 170,303 * e^{0+0.45^2/2}\ =\ \$188,449.$$

---

[4]   All costs at this point are in a base year, which will likely be whatever year the costs were captured and reported, as on the constants tab. The base year in which the costs were captured in the constants tab should be labeled for the user's awareness. Ideally, these costs will be updated at least annually.

That is, the risk adjusted estimate for our example assuming "very high" risk is $188,449 (base year 2021 dollars).

Using the lognormal standard deviation equation, we can calculate the standard deviation of the mean estimate for our example:

$$lognormal\ std\ deviation = \sqrt{e^{2(0)+0.45^2}(e^{0.45^2} - 1)*(170,303)^2} = \$89,202.$$

The next set of cost inputs calculates the sustainment costs for each cyber weapon element. Sustainment is calculated monthly as a function of the enhancement cost and a sustainment cost factor. The sustainment cost factor is a user input on the constants worksheet. As with the enhancement cost, a point estimate and log-normal mean are calculated for the monthly sustainment cost. The sustainment equation is

$$Monthly\ Sustainment\ Cost\ =\ Enhancement\ Cost \times (Sustainment\ Cost\ Factor/12),$$

where the sustainment cost factor is defined on the constants worksheet and divided by 12 to convert the amount to a monthly cost. This monthly sustainment cost equation is used for both the sustainment point estimate and the sustainment mean estimate by using the appropriate enhancement cost in the equation. The monthly sustainment standard deviation uses the log-normal standard deviation equation defined earlier using the same sigma from the cost risk selection or risk override. Therefore, the model requires the same enhancement and sustainment cost risk factor for each individual element. That is, if the user selects a "very high" cost risk for a given element, the sigma value associated with "very high" is applied both to the enhancement and sustainment estimate. The user cannot choose a "very high" cost risk for a given element enhancement and then a "low" cost risk for that same element's sustainment.

All cost fields are automatically calculated based on the user inputs to the requirement characterization inputs and user inputs in the constants worksheet, which is described later in this chapter.

Time-Phasing

The final section of the acquisition plan worksheet is for the phasing inputs. This section automatically calculates several time-phasing—or schedule—inputs, while also allowing the user to define others. The first field in the phasing inputs section is the development start date for each element. This is a month-year input that the user enters directly into the acquisition plan inputs worksheet. The remaining time-phasing fields are automatically calculated on the worksheet. After the development start date is defined, the development duration is determined in the next field. The development duration is determined using the same method as the point estimate for cost. The development duration estimate in months draws from a lookup table on the constants worksheet that captures the average development duration for each effect type, target environment, and complexity combination. The next set of fields captures schedule data for operations. Once the development start date and development duration are defined for each effect, the operation start date is automatically calculated to begin the month following development.

The Time to Discovery field captures the expected duration between the operational start date and the date of vulnerability discovery in months for each line item. It is a user-defined input.[5] It is assumed that the element has a probability of usability equal to 1 during this period. This input is also a factor in the total O&M cost for a line item with the Ops Duration input multiplied by the monthly O&M cost estimate. The discovery date field is then automatically calculated based on the operational start date and the time to discovery. The final field, operations end, is then automatically calculated based on several inputs. This field incorporates the life of the effect beyond the vulnerability discovery date as the probability of usefulness declines over time and adversaries defend against the vulnerability. It is automatically calculated based on the user's ADL selection and the user-assigned parameters of the decay function for the high and low ADL options.

Constants

The constants worksheet captures the final set of cost framework inputs. Four sets of user-defined inputs ultimately affect the model's outputs:

- enhancement point estimates
- development durations
- sustainment cost factor
- ADL decay functions.

We briefly discussed each of these inputs the "Acquisition Plans" section of this chapter in the context of how they are used to automatically calculate cost and development schedules.

Weapon Element Point Estimates

The first set of user-defined constants consists of a table of point estimates. This table includes 108 point estimates based on every combination of the three cyber weapon element types, 12 target environments, and three complexity levels, as shown in Figure 2.3.

As a proof of concept, our initial model uses publicly available cost data for the procurement of zero-day exploits. Zerodium provides price ceilings for the bounty payout to developers of zero-day exploits, given certain characteristics (e.g., mobile versus desktop and type of

---

5   The cost model does not explicitly make the distinction between cyberweapon elements for zero-day and n-day vulnerabilities. The user can however include cyberweapons targeting n-day vulnerabilities by inputting 0 months in the "time to discovery" field, which will immediately introduce effectiveness decay according to the ADL chosen in the ADL field.

operating system).[6] We use these price ceilings to represent a starting point for a procurement cost to the program. We present low, medium, and high costs based on the further breakouts in the Zerodium data (e.g., different client software, email servers, types of messenger services, research techniques, web browsers, web servers, and web apps). In addition, there are labor costs to operationalize the procured exploit. Implementation labor costs would include such activities as integration, testing, documentation, and program management. The total cost of an exploit in the model framework is therefore represented by the following equation:

$$Total\ Exploit\ Cost\ =\ Procurement\ Cost + Operationalization\ Cost.$$

To estimate the operationalization cost, we used data suggesting that, on average, 58 percent of resources are used for finding the vulnerability while 42 percent of the cost is to develop an operational exploit (Ablon and Bogart, 2017).[7] Equivalently, our assumption is that 58 percent of the total exploit cost is for procuring the rights to the exploit, while the remaining 42 percent accounts for operationalizing it. Given these percentages and the Zerodium procurement costs, we estimated the cost to operationalize an exploit using the following equation:

$$Operationalization\ Cost = \frac{Procurement\ Cost * 0.42}{0.58},$$

where the procurement cost is the Zerodium bounty payout.

Table 2.1 summarizes the total exploit costs used in this cost model for various operating systems on both desktops and mobile devices. To provide an example of the cost calculations, the assumed high cost of a *Desktop–Windows* exploit includes the cost of procurement, $1 million from the Zerodium bounty payout data, and the operationalization cost of $724,000 using the operationalization cost equation. See the appendix for more information on exploit costs.

Development Durations

The second set of user-defined constants consists of the development duration estimates in months. This table on the constants worksheet mirrors the point-estimate table and includes

---

[6] It is important to note that Zerodium quotes bounty payout ceilings. The cost to an organization procuring the rights to a zero-day exploit will need to include some additional fee to account for a Zerodium profit margin.

[7] Ablon and Bogart, 2017, provides averages on the number of days to find the vulnerability and to develop the vulnerability into an operational exploit. The average number of days required for these two broad categories is how the percentage split between the activities was derived. We make some assumptions when applying this split to the Zerodium cost data. First, we use the time to find the vulnerability as a proxy for the cost to procure an exploitable vulnerability. This assumes the resources that would be incurred to find an exploitable vulnerability would be roughly equivalent to the procurement cost of an exploitable vulnerability. Second, while not explicitly stated, we believe Ablon and Bogart, 2017, does normalize working days to full-time-equivalent working days because they convert days to costs using average burdened labor rates.

TABLE 2.1

Exploit Costs

Exploit Target Environment	Total Exploit Cost ($000)		
	Low	Medium	High
Desktop–Windows	17.2	170.3	1,724.1
Desktop–macOS	17.2	98.0	344.8
Desktop–Linux	17.2	103.6	862.1
Desktop–OS agnostic	17.2	133.1	1,724.1
Mobile–Android	25.9	593.6	3,448.3
Mobile–iOS	25.9	432.8	4,310.3
Mobile–OS agnostic	25.9	519.3	4,310.3

SOURCE: Zerodium bounty payout data adjusted to account for operationalizing the exploit.

NOTE: Amounts in thousands of base year 2021 dollars.

a development duration estimate for each of the same 108 combinations defined by element type, target environment, and complexity. Similar to the point estimates, the development durations are based on historical data and subject-matter expert (SME) estimates.

Sustainment Cost Factor

The third user-defined input on the constants worksheet is the sustainment factor. The factor is defined as a percentage in the model and is applied to the enhancement cost for each effect. Therefore, the higher the enhancement cost, the higher the assumed sustainment cost. This is common in software sustainment estimates. The current model does not vary the sustainment factor and applies the same, single factor to all elements. The sustainment factor is based on SME input.

Adversary Defense Level Decay Functions

The final user-defined input is the parameterization for the decay functions used in modeling element decay after a vulnerability is discovered. The user enters the decay function parameter for the high and low ADLs. Figure 2.4 depicts some notional decay functions with the associated exponential coefficients used in each function.

The figure shows how the usefulness of the element decays over time from the date of discovery. For example, defining a low ADL with a –0.1 coefficient in the function's exponent (the yellow line) assumes that the effect is 60-percent useful five months after discovery and 30 percent useful 12 months after discovery. As the value of the coefficient decreases below zero, the curves get steeper, and usefulness decays more quickly. In these notional curves, the –0.75 coefficient example decays the quickest, with the cyber weapon essentially useless six months after discovery.

FIGURE 2.4

Notional Operational Month After Discovery

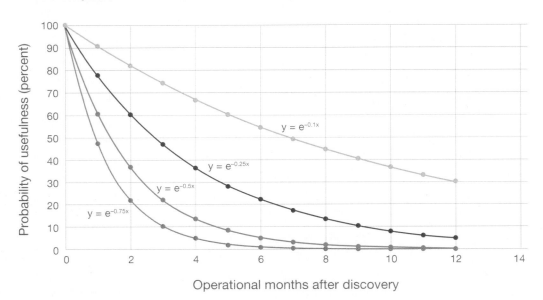

Other Factors

The remaining set of factors is not intended for the user to adjust in the short run. For instance, there is a set of escalation rates on the constants worksheet that applies inflation to the base-year or current-year estimates. These would likely be changed annually as new inflation indexes are released. The model uses the 2021 Navy and Marine Corps inflation rates (Naval Center for Cost Analysis, 2021).

Finally, the constants worksheet includes a lookup table for the cost risk scales and associated statistics that are used in the mean and standard deviation estimates discussed in "Acquisition Plans" section of this chapter.

Outputs

In this section, we describe the outputs of the cost-estimating framework, including production demand (or development demand signal), the cost of the JCW program, and a budgeting obligation authority CDF.

Production

The production portion of the cost model summarizes the cyber weapon elements and their associated high-level schedules. The schedule outputs provide the monthly status for each element that is listed in the acquisition plans. Each element can take on one of three monthly statuses:

1. **acquisition month:** a month in which the element is in development and not yet operational
2. **sustainment month:** a month in which the element is in operational use and being maintained
3. **neither in development nor operational:** a month either prior to development beginning for an element or a month after an effect is no longer operational.

Figure 2.5 depicts a notional output of ten cyber weapon elements.

The item number, type, target environment, ADL, acquisition start, operations start, and operations end are all based on the user inputs on the acquisition plan inputs worksheet. The model then displays the high-level schedule for each of the items in the element list. As the figure shows, each month that is an acquisition month for a given item is highlighted yellow, and each month that is a sustainment month for a given item is highlighted green. Months in which the element is not yet in development or operational are blank (or white).

Note that the sustainment months include a number. We note in the spreadsheet that this number reflects the probability that the element is useful in a given month. So, for instance, item 1 in Figure 2.5 is demonstrating decay after the first 12 months. Between the operations start date and the discovery date—in this example, from December 2021 to the end of November 2022—there is a 1.0 probability of usefulness, i.e., $P(U) = 1.0$. We assume that, until the vulnerability is discovered, the element is 100-percent useful. After the date of discovery, we begin to note that $P(U) < 1$ as adversaries begin to adopt defenses against discovered vulnerabilities. The decay of element usefulness demonstrated in the requirement outputs is a direct result of how the user parameterized the decay function on the constants worksheet and the selection of ADL on the acquisition plan inputs worksheet.

Another function of the production portion of the model is its linkage to the requirements worksheet. The number of elements that are operational in each month is calculated on the requirements output worksheet. This count is broken down by effect type, giving the total operational exploits, implants, and payloads in each month. The linkage to the requirements inputs is evident with conditional formatting by highlighting whether outputs meet input requirements. If the operational count for each element type meets the requirement objective input on the requirements input worksheet, then the count will be highlighted dark green. If the operational count for each element type meets the requirement threshold input on the requirements worksheet, then the count will be highlighted light green. Finally, if the operational count for each element type does not meet the threshold requirement, the count will be highlighted in red. The top portion of Figure 2.5 demonstrates this schema. In this example, all outputs meet the operational objective; therefore, all the operational count cells are highlighted dark green. This provides a quick visual for the user to identify anticipated requirement gaps.

FIGURE 2.5
Requirements Output

Legend:
- Does not meet Threshold
- Meets Threshold
- Meets Objective

Operational Exploit Count	2	2	2	9	9	9	9	11	11	12	12	13	12	12	12	12	12	14	15	18	18	19	12	11	11	12	12	12	12	12	9	9	9	2	2	2	15	16	16	16	17	16	15
Operational Implant Count	0	0	0	0	0	0	0	0	0	0	0	0	0	0	0	0	0	0	0	0	0	0	0	0	0	0	0	0	0	0	0	0	0	0	0	0	0	0	0	0	0	0	0
Operational Payload Count	0	0	0	0	0	0	0	0	0	0	0	0	0	0	0	0	0	0	0	0	0	0	0	0	0	0	0	0	0	0	0	0	0	0	0	0	0	0	0	0	0	0	0

Legend:
- ▨ = Acquisition Month
- P(U) = Sustainment Month

where P(U) = probability that the cyber weapon element is useful in a given month; P(U) drops after vulnerability discovery date

Item #	Cyber Weapon Element	Target Environment	Adversary Defense Level	Acq Start	Ops Start	Discovery Date	Ops End
1	Exploit	Desktop - Windows	High	Dec-20	Dec-21	Dec-22	May-23
2	Exploit	Desktop - macOS	High	Dec-20	Dec-21	Dec-22	May-23
3	Exploit	Desktop - Linux	High	Dec-20	Dec-21	Dec-22	May-23
4	Exploit	Desktop - Any OS	High	Dec-20	Aug-22	Aug-23	Jan-24
5	Exploit	Desktop - Any OS	High	Dec-20	Dec-21	Dec-22	May-23
6	Exploit	Desktop - Any OS	High	Dec-20	Aug-21	Aug-22	Jan-23
7	Exploit	Mobile - iOS	Low	Dec-20	Dec-21	Dec-22	Nov-23
8	Exploit	Mobile - Android	Low	Dec-20	Dec-21	Dec-22	Nov-23
9	Exploit	Mobile - Any OS	Low	Dec-20	Dec-21	Dec-22	Nov-23
10	Exploit	Mobile - Any OS	Low	Dec-20	Aug-22	Aug-23	Jul-24

The monthly P(U) matrix spans Dec-20 through Jul-23, grouped by fiscal year (FY 2021: Jan-21–Dec-21; FY 2022: Jan-22–Dec-22; FY 2023: Jan-23–Jul-23).

Selected P(U) values by row (operational period through discovery decay):

- Row 1: 1.0 (Dec-21 … Dec-22) then 0.6, 0.4, 0.2, 0.1, 0.1 (Jan-23–Apr-23)
- Row 2: 1.0 … 0.6, 0.4, 0.2, 0.1, 0.1
- Row 3: 1.0 … 0.6, 0.4, 0.2, 0.1, 0.1
- Row 4: 1.0 (Aug-22 …) then 1.0, 1.0 (May-23–Jul-23)
- Row 5: 1.0 … 0.6, 0.4, 0.2, 0.1, 0.1
- Row 6: 1.0 (Aug-21 …) then 0.6, 0.4, 0.2, 0.1, 0.1 (Aug-22–Dec-22)
- Row 7: 1.0 … 0.8, 0.6, 0.5, 0.4, 0.3, 0.1
- Row 8: 1.0 … 0.8, 0.6, 0.5, 0.4, 0.3, 0.1
- Row 9: 1.0 … 0.8, 0.6, 0.5, 0.4, 0.3, 0.1
- Row 10: 1.0 (Aug-22 …) 1.0, 1.0 (May-23–Jul-23)

NOTE: This is a partial snapshot of the requirements output from a much larger worksheet in the Excel model.

22

Costs

The cost portion of the model framework provides various cost output summaries for the JCW program based on all user inputs on the acquisition plans and constants worksheets. The costs are time-phased by FY on each cost summary. The FY time-phasing uses the acquisition start date, operational start date, and operational end date for each item to apportion acquisition and O&M costs to each FY. For instance, if an element's development spans multiple FYs, the amount apportioned to each FY for each element is determined by the ratio of the number of development months in each FY divided by the total number of development months for the element. This calculation determines the percentage of cost attributed to each FY. As an example, if an element takes 12 months to develop and the development start date is July 1, 2021, three of the 12 development months will be in FY 2021 and nine will be in FY 2022. The model would therefore apportion three out of 12 (or 25 percent) of the element development cost to FY 2021 and nine out of 12 (or 75 percent) of the element development cost to FY 2022. For O&M costs, the framework simply multiplies the number of operational months by the monthly sustainment cost, as calculated on the acquisition plans worksheet. The apportionment of O&M costs to a given FY is based on the number of months each element is operational in the FY.

The costs mirror the same list of items on the acquisition plans and requirements worksheets. Figure 2.6 depicts a notional cost summary output for the first ten items. The item number, element type, and target environment information for each element are listed in the cost outputs worksheet. Included in the cost outputs are time-phased summaries like that depicted in Figure 2.6 for the enhancement (or development) cost and the sustainment cost. Additionally, there are outputs based on both the point estimates and the mean estimates.[8]

Included in the cost outputs worksheet are bar charts depicting the costs in graphical form. Figure 2.7 shows a notional cost output.

This example includes the enhancement and sustainment mean costs in a single bar chart. The cost model framework provides similar bar charts for enhancement and sustainment separately (both their point estimates and mean estimates). All costs in the cost outputs worksheet are presented in base-year dollars.[9] The user can find a summary of then-year (TY) costs (i.e., costs adjusted for inflation) on the obligation authority CDF worksheet discussed in the next section.

Obligation Authority Cumulative Distribution Function

The obligation authority CDF portion of the cost framework provides two key outputs: a tabular and graphical depiction of the uncertainty around cost estimates and the TY costs,

---

[8] See the discussion in the "Element Costs" section for the difference between the point estimate and the mean estimate.

[9] In the current model, the base year is set to FY 2021. This can be updated annually to align with the most current base year.

FIGURE 2.6

Cost Output Summary

Item #	Cyber Weapon Element	Target Environment	Enhancement Cost by Fiscal Year (Point Est) 2020	2021	2022	Sustainment Cost by Fiscal Year (Point Est) 2020	2021	2022
		Total	$ 3,346,050	$ 9,249,848	$ 7,484,812	$ -	$ 435,036	$ 3,506,969
1	Exploit	Desktop - Windows	$ -	$ 141,919	$ 28,384	$ -	$ -	$ 49,672
2	Exploit	Desktop - macOS	$ -	$ 81,670	$ 16,334	$ -	$ -	$ 28,584
3	Exploit	Desktop - Linux	$ -	$ 86,328	$ 17,266	$ -	$ -	$ 30,215
4	Exploit	Desktop - Any OS	$ -	$ 862,069	$ 862,069	$ -	$ -	$ 100,575
5	Exploit	Desktop - Any OS	$ -	$ 110,897	$ 22,179	$ -	$ -	$ 38,814
6	Exploit	Desktop - Any OS	$ -	$ 17,241	$ -	$ -	$ 1,006	$ 6,034
7	Exploit	Mobile - iOS	$ -	$ 494,685	$ 98,937	$ -	$ -	$ 173,140
8	Exploit	Mobile - Android	$ -	$ 360,665	$ 72,133	$ -	$ -	$ 126,233
9	Exploit	Mobile - Any OS	$ -	$ 432,787	$ 86,557	$ -	$ -	$ 151,475
10	Exploit	Mobile - Any OS	$ -	$ 2,155,172	$ 2,155,172	$ -	$ -	$ 251,437

NOTES: This is a partial snapshot of the cost summary output. The totals do not sum the items in the snapshot because the list is much longer and includes many more items and FYs.

FIGURE 2.7

Cost Output Summary Bar Chart

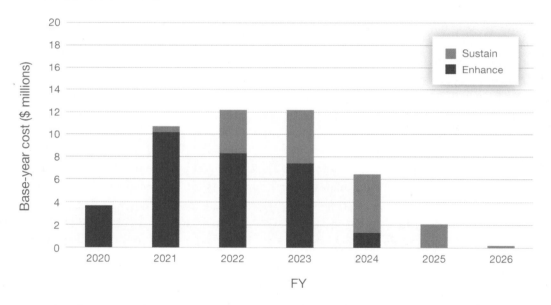

or escalated costs in addition to the base-year costs. As the title of the worksheet suggests, this output is most useful to the user for obligation authority budgeting. Obligation authority budgets are always in TY dollars, and, in some cases, budgets can be authorized at levels above the mean estimate to account for risk.

Because the basic cost framework is constrained to rely only on the use of the native capability and functions in Excel (i.e., without the use of Excel add-ins, macros, or Visual Basic for Applications code), we avoided the more-common cost uncertainty approaches. The typical approach to analyzing cost uncertainty is to use a Monte Carlo method to understand the overall uncertainty based on component distribution. Although this method is generally easier for an analyst to implement, it typically requires a plug-in or code to implement in Excel. An equally acceptable but less common method is the method of moments (MoM) approach (Naval Center for Cost Analysis, 2014, p. 80). It is not commonly used because it requires the analyst to derive somewhat complex formulas (partial differentials) to calculate the moments of the distribution. For complex cost models, this approach becomes too time-consuming. However, the budgeting problem that this tool addresses is a specific case of the more generalizable portfolio problem where semicorrelated items are added together to determine the total value. The underlying math of the MoM approach greatly simplifies these cases. To slightly oversimplify, the means and variances of the component distributions add to the means and variance of the final distribution.[10] Another advantage of the MoM approach is that it is not a simulation but, rather, a deterministic calculation. So, a spreadsheet model can implement the MoM approach very quickly.

The uncertainty modeling in the tool assumes that the input and output uncertainty distributions are log-normal.[11] This assumption is based on our prior research on defense program cost growth (Arena et al., 2006). An advantage of the log-normal distribution is that it is well understood and characterized. We used the formulas in the *Joint Agency Cost Schedule Risk and Uncertainty Handbook* to calculate the various characteristics for the distribution (Naval Center for Cost Analysis, 2014, p. A-17).

Example Implementation

To get a sense of how the MoM approach compares with the more traditional Monte Carlo method to estimate uncertainty, we created and analyzed a simple example both ways using the same assumptions. We created three line items with a point value of $5 with high uncertainty. The items were uncorrelated. The MoM output derives from this spreadsheet, but we implemented the Monte Carlo in an analogous model in Analytica. Figure 2.8 shows the CDF for the MoM approach versus the points from the Monte Carlo approach. The values agree to within 0.5 percent, suggesting that the MoM approach is reliable for a budget uncertainty calculation.

Optionally, the tool also allows the analyst to add an uncertainty correlation between items using a correlation coefficient. This pairwise correlation is applied to all items appropriated or obligated *within* a single FY. The correlation does not apply across FYs and budget accounts. The default value for the correlation coefficient is zero. The *Joint Agency Cost Schedule Risk and Uncertainty Handbook* recommends a correlation default of 0.3, but that is for the correlation of elements within a single estimate (Naval Center for Cost Analysis, 2014, p. 48). It is unclear what the appropriate value is for a portfolio of estimates.

---

[10] See Bevington, 1969, for more details on the propagation-of-errors method (which is the same approach as MoM) and how it accounts for correlation.

[11] Note that these assumptions are not required of the MoM approach. But translating the resulting mean and variance values into CDF does require assuming a distribution.

FIGURE 2.8

Comparing the MoM and Monte Carlo Approaches

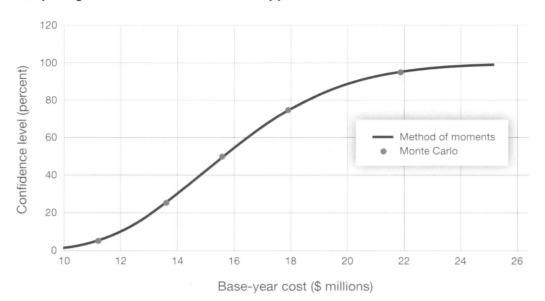

The costs in the obligation authority CDF are rolled-up total costs of all items in the acquisition plan for a given year. The user can select any year within the Future Years Defense Program to display the obligation authority CDF estimate for that year. The user has the option to display the enhancement or sustainment costs CDF. Figure 2.9 shows a notional CDF output for FY 2024 enhancement costs in TY dollars.

The model shows the median, mean, and 80-percent confidence level estimate for the options selected, those options being the FY and activity type (e.g., enhancement or sustainment). In this notional example, the median or, equivalently, the 50-percent confidence level is slightly below $1.5 million, while the mean or expected value is slightly above $1.5 million (TY dollars). The 80-percent confidence level in this notional example is approximately $2.2 million (TY dollars). Interpreting these values means that, if the organization wishes to be 80-percent confident that it will budget enough funds, given the uncertainty in the inputs, it should budget for $2.2 million (TY dollars) in FY 2024 for enhancement activities. Again, details on risk and uncertainty and how they are calculated at the item level are discussed in the "Acquisition Plans" section earlier in this chapter.

In addition to the graphical CDF in Figure 2.9, the user can view a tabular form of the confidence levels based on the same user selections for FY and activity (e.g., enhancement or sustainment). Figure 2.10 shows the notional table that accompanies the same notional example in Figure 2.9. This can give the user a list of base-year and TY costs at several different confidence levels.

FIGURE 2.9

Example Obligation Authority CDF

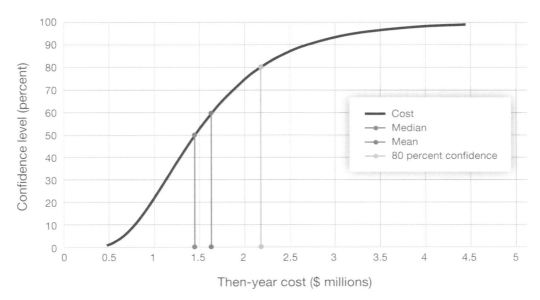

FIGURE 2.10

Confidence Level Outputs

CDF (Base Year $)	Cost (Then Year $)	Confidence Level
$379,009	$471,670	1%
$432,266	$537,948	2%
$526,506	$655,228	5%
$627,343	$780,718	10%
$706,075	$878,698	15%
$775,637	$965,267	20%
$840,752	$1,046,301	25%
$966,587	$1,202,900	35%
$1,095,550	$1,363,393	45%
$1,164,005	$1,448,584	50%
$1,236,737	$1,539,098	55%
$1,401,744	$1,744,447	65%
$1,611,543	$2,005,537	75%
$1,746,833	$2,173,903	80%
$1,918,930	$2,388,075	85%
$2,159,755	$2,687,778	90%
$2,573,394	$3,202,545	95%
$3,134,430	$3,900,745	98%
$3,574,873	$4,448,868	99%

Summary

In this chapter, we proposed a framework for estimating the costs of the JCW program through a combination of inputs designed to reflect key parts of the program: requirements, acquisition plans, cyber weapon element costs, time-phasing (schedule), and various non–

user-defined constants. The framework is anchored in requirements that reflect operational needs in terms of a quantity of cyber weapon elements. Accounting for the numerous uncertainties, the framework predicts the number of cyber weapon elements needed in production to meet the operational objectives. Once that production demand signal is established, it is possible to estimate the necessary costs and the obligation authority.

We implemented the framework in Microsoft Excel so that the sponsor could readily make use of it.

CHAPTER THREE

Exploratory Model and Simulation

Chapter Two described a cost-estimating framework implemented as an Excel tool that can be used to calculate acquisition and operations costs and timelines for a portfolio of JCW program capabilities. It also identified several challenges related to JCW cost estimation, including the significant uncertainty in terms of demand signal and vulnerability life span.

To capture the impact of these uncertainties, we developed an exploratory cost model and simulation using R.[1] The model is a direct implementation of the Excel model described in Chapter Two but layers a simulation component atop that model to explore the impact of assumptions related to the life span of a vulnerability and the weapons developed to exploit the vulnerability, which we collectively refer to *exploits* in this chapter.[2]

Simulation Approach

The exploratory model has two main components: a model that implements the cost-estimating framework described in Chapter Two, and a simulation that wraps around it. The model implements all the relevant pieces from the Excel model: the requirements over time and an anticipated set of projects, including their timelines. It performs many of the calculations in exactly the same way as the Excel model, except when there is significant potential uncertainty (e.g., acquisition timeline, operations timeline, cost). In those cases, instead of using a specific value from a potential distribution of outcomes, the model makes a random draw over the specified distribution. The simulation component then wraps around the model, running each of the portfolios of capabilities a user-specified number of times so that the uncertainty in both costs and timelines can be explored.

We described how a single case is handled in this exploratory model, but the user will typically specify many cases to be run simultaneously. The distinction between different cases can be based on changes to any of the model inputs, but the main cases explored in this chapter focus on changing the potential set of requirements, the ADL, and the number of capabili-

---

[1] See the R Project for Statistical Programming, undated, for a more detailed description of the R programming language.

[2] Theoretically, the model is not limited to exploits, but there are limited data on payloads and implants. Thus, the example in this chapter focuses on exploits.

ties being purchased. The R model also allows the user to automate increasing the portfolio of capabilities, which can help quickly assess the exploit acquisition plans which will satisfy requirements under varying model assumptions.

Features

The R exploratory model adds a few key features on top of what the Excel model provides:

1. **The exploratory model allows multiple cases to be run simultaneously.** Unlike the goal of the Excel model, which may be to understand the cost and timeline effects of a particular portfolio, the exploratory model is fundamentally designed to facilitate comparisons across sets of cases and examine a tradespace. As an example, the model is set up to compare across different levels of requirements.

2. **The exploratory model allows the user to compare across sets of assumptions.** As described later in this chapter, there are not just distributions of potential outcomes for various values in the model; different datasets suggest that there are entirely different potential distribution ranges for some individual input values. See the appendix for two methods of estimating the life span of an exploit—using the last major or minor release as the assumed introduction date—with resulting distributions that are significantly different. The model can simulate multiple options to capture the effect on the overall portfolio of exploits required.

3. **The exploratory model allows the user to understand the interactions between multiple distributions of outcomes.** Any time there are multiple distributions in an analytic problem, interaction effects can lead to outcomes outside the obvious expected ranges. Simulating each case hundreds or thousands of times alleviates these concerns and allows the entire distribution of outcomes to be understood.

4. **The exploratory model can model duplicates of the potential capability profile automatically.** Changes to the requirements or operational timelines of capabilities will necessarily mean that more or fewer capabilities are required. The exploratory model handles that automatically, by replicating the baseline set of capabilities a user-specified number of times. This allows the user to compare, for example, sets of investments that both meet the same requirement even if they have significantly different life-span distributions.

Limitations

In addition to these key features, there are some important limitations to keep in mind when using this analytic approach:

1. **There is still a large data requirement.** Although the exploratory model improves understanding of the tradespace, useful modeling outcomes still require some

knowledge of the eventual values for hard-to-define parameters (e.g., the longevity of exploits).

2. **Exploration can help determine what parameters are important, but users should take care when drawing insights about specific values.** Results from the exploratory model might suggest, for example, that requirement levels have more of an effect on the total cost and ability of the JCW program to meet its mission. However, the model still cannot define what the appropriate requirement levels may be for meeting program objectives.

Example Analysis

In this section, we present a notional analysis designed to demonstrate and explore the capability of the exploratory modeling and simulation approach. For this exploratory analysis, we researched input variables and tried to bookend them when possible.[3]

Assumptions

To ensure clarity, we avoided changing too many parameters simultaneously. In the example analysis that follows, we standardized a few parameters that the exploratory model can vary. First, we supplied the model with a single, notional initial acquisition plan that consisted of ten exploits for each target environment (desktop, mobile) over a simulation period of five years. The model replicates this baseline acquisition plan to identify the acquisition plan that satisfies the requirements under the other assumptions.

Table 3.1 describes the simulated acquisition times used in this analysis. The acquisition time for each exploit is simulated based on a specified complexity attribute. We set the baseline input so that 60 percent of the exploits were categorized as high complexity, with the remaining 40 percent set to medium complexity, like the setup of the Excel model, which contains a mix of medium- and high-complexity exploits. The proportional split is an assumption that slightly more high-complexity exploits will be acquired, but the split may be parameterized in future analyses. The simulation generates an acquisition time from a normal distribution characterized by the means and standard deviations shown in the table. The mean acquisition times used for the three levels of complexity align with the values used in the Excel model discussed in Chapter Two.[4]

Table 3.2 describes the set of normal distributions representing the operational time of each exploit. These ranges of potential operational times are informed by an analysis of the

---

[3] The appendix details our research on the life span of various vulnerabilities using publicly available data sources.

[4] We did not explore standard deviations in the Excel version of the model. In this analysis, we set the standard deviation for acquisition time as the nearest whole number to one-sixth of the mean value used.

TABLE 3.1

Simulated Acquisition Times

Complexity	Mean Acquisition Time	Standard Deviation
High	18	3
Medium	9	2
Low	6	1

NOTES: Mean acquisition times and standard deviations in months are used for normal distributions representing possible acquisition times in the model. Acquisition times for simulated exploits were generated based on the complexity attribute.

TABLE 3.2

Simulated Operational Times

Target Environment	ADL	Mean Operational Time	Standard Deviation
Desktop	High	5	2
Desktop	Medium	14	5
Desktop	Low	22	7
Mobile	High	3	1
Mobile	Medium	5	2
Mobile	Low	6	2

NOTES: Mean operational times and standard deviations in months used for normal distributions representing possible operational times within the model. Operational times for simulated exploits were generated based on combinations of target environments and ADL.

Google Project Zero dataset (Google, 2021). The data contain patch dates for 187 detected zero-day exploits ranging back to 2014. As a way of estimating the operational times of these exploits, we captured the length of time between the last major update of a software and the date of the patch. Using this method, we calculated operational times for 133 of the entries in the dataset. We then classified each entry into the broader target environment categories of desktop ($n = 112$), mobile ($n = 19$), and infrastructure ($n = 2$) used within the model. Because there were only two infrastructure entries, this exploratory analysis focused on desktop and mobile exploits. The mean operational time, particularly when a vulnerability is readily identified and patched, was quite short, at three to five months. For more detail on our analysis of the Project Zero data, see the appendix.

Figure 3.1 shows survival curves for combinations of target environments and ADLs, reflective of the normal distributions representing possible operational times as described

FIGURE 3.1

Simulated Operational Times, by Adversary Defense Level

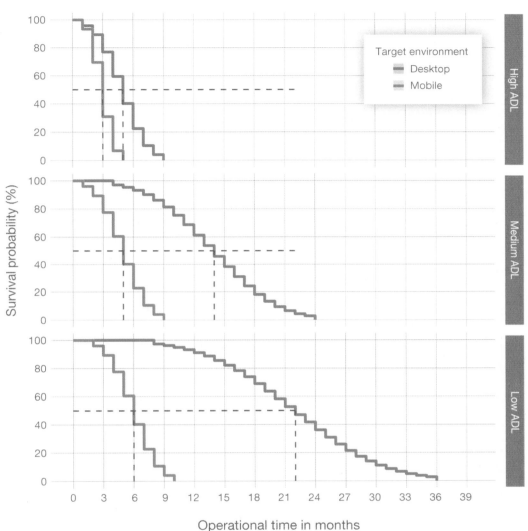

in Table 3.3. The curves show the percentage chance of an exploit surviving as a function of the operational time in months. Dotted lines represent the midpoint values, where one-half of the exploits are expected to survive longer than the operational time intersecting with a 50-percent survival probability and one-half are expected to survive for a shorter amount of time. For example, in cases with a low ADL, mobile exploits may survive up to approximately ten months, with a midpoint 50-percent survival probability of six months. For desktop exploits under the same ADL assumption, operational times may extend up to 36 months, with a midpoint 50-percent survival probability of 22 months. The simulation randomly

TABLE 3.3

The Design of the Experiments

Case	ADL	Requirement Level	Threshold	Objective
1	High	Low	1	2
2	Medium	Low	1	2
3	Low	Low	1	2
4	High	High	5	10
5	Medium	High	5	10
6	Low	High	5	10

draws values from the appropriate curve to determine the operational life span of a given exploit in a particular Monte Carlo iteration.

Design of Experiments

Table 3.3 describes this example exploratory analysis.[5] The Case column assigns numbers to the illustrative, simulated scenarios created for this experiment. AD is a proxy for how advanced the adversary is and, consequently, the life span of the exploit, as described in Figure 3.1. Requirement Level represents the expected operational requirement: We modeled low and high requirements in this design. In the low requirement–level cases, we set a threshold of one and an objective of two for each target environment. High requirement–level cases had threshold and objective values of five and ten, respectively, for each target environment.[6]

Results

Figure 3.2 shows the baseline results for case 1, which included a high ADL and low requirements, as described in Table 3.3. The x-axis represents time, and the y-axis is the number of available operational exploits (on average across the Monte Carlo simulation runs). The lower horizontal line is the threshold operational requirement, while the higher horizontal line represents the objective operational requirement. As Figure 3.2 shows, although the notional initial acquisition of ten exploits was purchased in each of the desktop and mobile cases, the high ADL is able to suppress them, such that the objective operational requirement is met for less than six months of the simulation and that the threshold operational requirement is met for approximately 24 months for desktop exploits across five FYs. Mobile exploits in this case

---

[5] As a reminder, the majority of the calculations for a given case are taken directly from the Excel model described in Chapter Two, so a user who is interested in the specifics of the calculations should look there for the details.

[6] Determining the appropriate value for these requirements was beyond the scope this project but would be a useful focus area for future work.

FIGURE 3.2

Baseline Results for Case 1: Low Requirement and High ADL

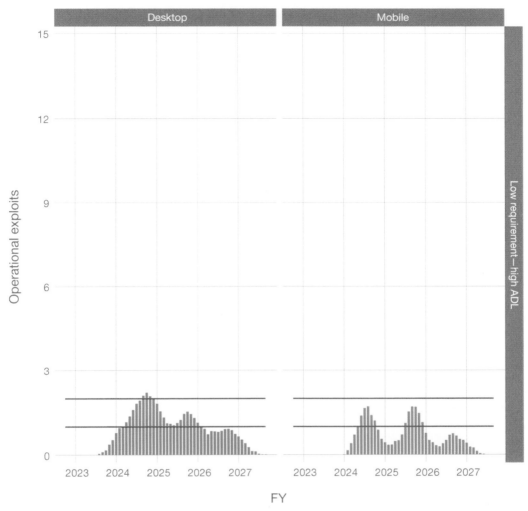

never meet the objective operational requirement, and they meet the threshold requirement in only two six-month periods.

Figure 3.3 expands Figure 3.2 to show the comparison across the first three cases in the experiment. The top row of two plots in Figure 3.3 matches the results described in Figure 3.2, while the second and third rows of plots show the effect of lowering the ADL. Unlike in the high ADL case, desktop exploits exceed objective requirements much of the time in cases with an assumed medium or low ADL. Although changing the ADL has a more limited effect on the mobile exploits, transitioning from the high ADL to the low ADL still results in the threshold requirement being achieved more often over the five-year simulation period. Going from high to medium or low ADL results in the objective requirement going from never

FIGURE 3.3

Baseline Results for Low Requirements in Cases 1–3

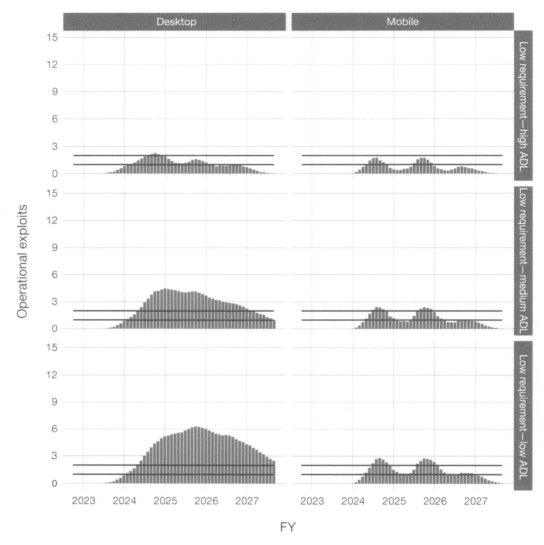

NOTES: Data shown correspond with cases 1–3 in Table 3.3. The requirement level is low (a threshold operational requirement of one and an objective operational requirement of two exploits for both desktop and mobile).

achieved to sometimes achieved. This suggests that, given these assumptions about operational life spans, the baseline portfolio of desktop exploits may be sufficient when the ADL is medium or low, while the portfolio of mobile exploits will likely need to be augmented in any of the cases, even at the low requirement level.

Figure 3.4 shows the results of increasing the portfolio of desktop exploits acquired in the three low-requirement scenarios. Moving across the figure from left to right, each portfolio

FIGURE 3.4

Duplications of Desktop Acquisition Portfolio for Low Requirement in Cases 1–3

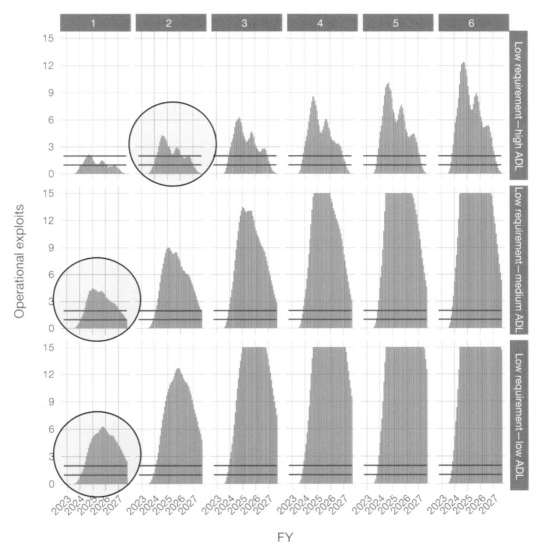

NOTES: Column 1 represents the baseline input of ten desktop exploits acquired over the five-year simulation period. Each column to the right subsequently adds an additional ten exploits to the acquisition plan.

contains an additional ten desktop exploits produced over the five-year simulation period. The first column of results contains the same baseline results shown in Figure 3.3. In these low-requirement cases, the baseline acquisition plan of ten desktop exploits can satisfy the operational objective requirement under an assumption of medium or low ADL. In the case with a high ADL, adding an additional ten desktop exploits to the acquisition plan can meet the objective requirement, as highlighted in the top portion of the second column in

Figure 3.4. In summary, for cases 1, 2, and 3, shown here, an acquisition plan containing 20, ten, and ten desktop exploits, respectively, will meet the operational requirement.

We conducted the same exercise to determine the mobile exploit acquisition plans that would meet objective requirements (see Figure 3.5). Moving left to right, each column, represents an additional ten mobile exploits added to the acquisition plan. In the top row (case 1, high ADL) the operational requirement is not consistently met until the chart highlighted

FIGURE 3.5

Duplications of Mobile Acquisition Portfolio for Low Requirement in Cases 1–3

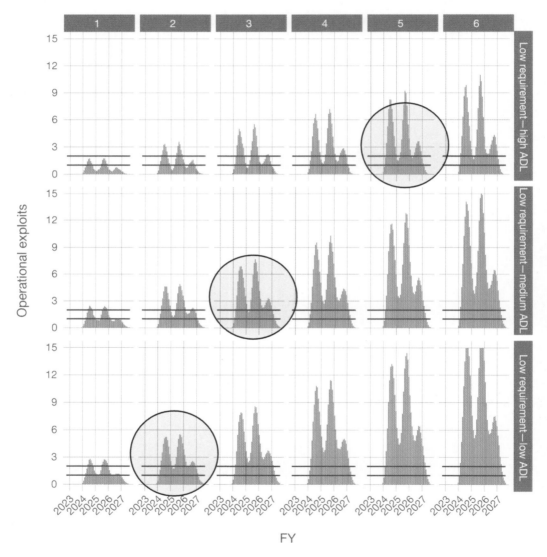

NOTES: Column 1 represents the baseline input of ten mobile exploits acquired over the five-year simulation period. Each column to the right subsequently adds an additional ten exploits to the acquisition plan.

in column 5, suggesting an acquisition plan containing 50 mobile exploits. Under a medium ADL, 30 mobile exploits will meet the objective, and 20 mobile exploits will meet the objective in the low-ADL case shown in the bottom row. This illustrates the impact that vulnerability and exploit life spans can have on program planning goals. Depending on the underlying assumptions, the number of exploits required to meet operational goals and the associated budget required to produce those exploits may have a large range.

Increasing the requirements in cases 3–6 can greatly change the acquisition plan solution identified from model outputs to meet operational goals. Figure 3.6 shows the tradespace of desktop acquisition plans in high requirement cases against the varied ADL. In this example, the baseline input of ten exploits is not able to meet the objective requirement under any assumption of ADL. Even in the sixth column, representing 60 desktop exploits in the acquisition plan, the objective operational requirement is met for only about six months of the simulated five-year period. In cases with a medium and low ADL, acquisition plans of 30 and 20 desktop exploits, respectively, will meet the objective operational requirement for extended periods while consistently meeting the threshold operational requirement.

Figure 3.7 explores acquisition plan options for mobile exploits of the same cases. Similar to the desktop exploits, in the cases shown in the top row with a high ADL, even the largest acquisition plan of 60 mobile exploits reaches the operational objective requirement only for brief periods of the simulated time frame. An acquisition plan of 50 mobile exploits will meet the objective in the case with a medium ADL during two separate periods of less than six months, while an acquisition plan containing 40 mobile exploits will have the same result for the low ADL case.

Combining the desktop and mobile acquisition levels needed to meet requirements as shown in Figures 3.3–3.7, we can produce solution acquisition plans for each case. The acquisition plans used to determine the costs shown in Figure 3.8 reflect the highlighted sections in the previous charts for each case. Consistent with the range in the number of exploits required to meet program goals, the resulting total costs vary widely across the cases. As may be expected, the most expensive case has a high requirement level and a high ADL, with an expected cost range from approximately $275 million to $290 million over five years. Comparatively, in the low-requirement case with high ADL, costs are expected to fall close to $200 million, approximately one-third less than the high-requirement case. Changing the assumption of ADL may also have a large effect on expected overall costs. For example, in the high-requirement case with a medium ADL, expected costs are approximately $220 million to 240 million. This illustrates that, although requirement levels are significant drivers of program cost, uncertainties, such as those around expected exploit life spans, may be equally influential.

FIGURE 3.6

Duplications of Desktop Acquisition Portfolios for High Requirements in Cases 4–6

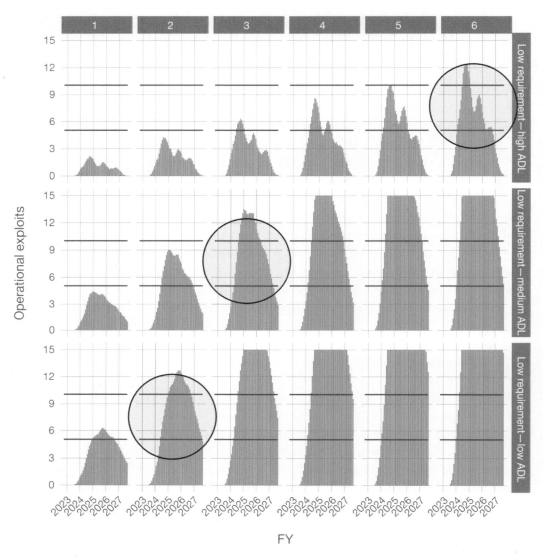

NOTES: Column 1 represents the baseline input of ten desktop exploits acquired over the five-year simulation period. Each column to the right subsequently adds an additional ten exploits to the acquisition plan.

Summary

Chapter Two described a cost-estimating framework that could be implemented to understand the potential cost and operational timeline requirements for a particular portfolio of cyber effects. This chapter has described a simulation model that took that cost-estimating

FIGURE 3.7

Duplications of Mobile Acquisition Portfolios for High Requirements in Cases 4–6

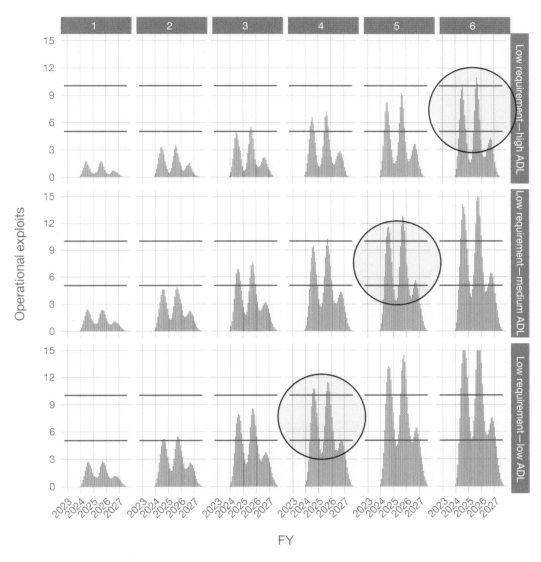

NOTES: Column 1 represents the baseline input of ten mobile exploits acquired over the five-year simulation period. Each column to the right subsequently adds an additional ten exploits to the acquisition plan.

framework and used it to explore the significant uncertainty in potential costs over a five-year period. To do this, we estimated the mean operational time of exploited vulnerabilities based on available data and found that timelines can be very short in worst-case conditions, approximately three to five months. Thus, even holding constant some parameters of concern in our example analysis—such as the simulated acquisition and operational timelines—we

FIGURE 3.8

Total Acquisition Plan Costs for All Cases

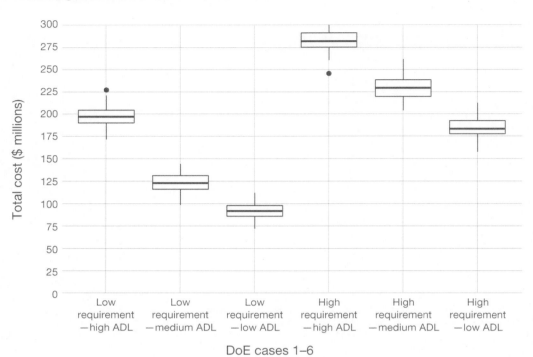

DoE cases 1–6

NOTE: Range of expected costs reflective of acquisition plans found to meet operational objective requirements across cases 1–6 presented in the design of experiments.

found that the JCW program can incur a large range of potential costs, depending on the operational requirements and ADLs, ranging from $90 million to $290 million.

In the example analysis presented, we used the best available data, but some of the appropriate distributions of values for various parameters remain unclear. It is likely that many of the uncertainties described here—timelines and requirements, in particular—require additional analysis to better simulate. Still, in using bookend values, we were able to illustrate the potential tradespace of outcomes. As those values are further refined, there will still be a need for this type of tradespace analysis. However, we suspect that the tradespace will be more wieldable. In presenting this initial example analysis, we hope to both highlight the vastness of the current uncertainty and show that a point solution may not properly describe the potential uncertainty and, therefore, the risk associated with any given portfolio of cyber capabilities.

CHAPTER FOUR

Conclusions and Next Steps

This report presents analyses to inform the new JCW program as it embarks on its SWP acquisition. In the process, we identified key terms of interest relating to vulnerabilities and their exploitation.

To increase transparency and decision quality in JCW's pending acquisition, we brought together operational capability, schedule, and risk into a JCW cost-estimating framework to improve understanding of the potential costs and budgeting considerations.

During that process, we identified potential risks to the program related to the uncertainty of the life span of vulnerabilities and the derived cyber weapon capabilities that should be considered. We assessed the life spans of 133 historical vulnerabilities and developed an exploratory simulation that implemented the cost framework to quantify uncertainty as a function of the program's cost. We found that mean life span can be quite short for mobile and desktop vulnerabilities (three to five months) where potential adversaries have a high defense level (i.e., an ability to rapidly identify and patch a vulnerability).

Drawing on the data and assumptions about operational demand, we found that there is significant uncertainty in the potential costs of the JCW program, from $90 million to $290 million over a five-year period.

It is important to note that the operational demand level is notional, as is the cost of procuring and operationalizing an exploit, and these costs could change as more data become available. Other parameters could also be explored, such as the complexity of exploits.

The cost-estimating framework presented in this report serves as a foundation for future incremental improvements as a better understanding of the challenges and additional historical data become available.

Future Model Enhancements

Over time, as more acquisition and execution data are accrued, we will be able to update the model and inputs to better predict future costs. This section provides some thoughts on where future iterations of the model could be improved, providing guidance on what information and data will be useful in supporting those improvements.

Incorporating Historical Data on Cyber Weapon Element Cost

The most crucial improvement will include additional historical data to inform the point estimates in the constants worksheet. Most of the current inputs are based on SME judgment. As with any acquisition effort, it is preferable to use actual historical data. Therefore, in future iterations of the model, we would replace SME estimates used in the various element type/target environment/complexity combination point estimates with actual execution data for both cost and schedule duration.

Although analogous data provide more-robust and -defensible cost and schedule estimates than SME judgment does, this is not necessarily the goal. Ideally, over several years and model iterations, we would be able to collect enough data to conduct statistical analyses and fit data to distributions. For instance, there could be cost and schedule duration distributions for each cyber weapon element, providing a more complete picture of risk and uncertainty. CDFs could be determined by running simulations using the various distribution parameters.

Increasing the Granularity of Cyber Weapon Element Estimates

Another potential improvement involves increasing the granularity of the estimates. The current model estimates the rolled-up cost of an element. That is, each element line-item development cost includes all activities associated with the development of the element, such as the research conducted before development, development, testing, integration, training, and anything else required to operationalize the element. Perhaps an analysis of lower-level data would demonstrate distinctions worth modeling in the cost framework. For instance, we might find that the testing of an exploit for mobile operating systems is much more costly than for desktops; therefore, it might be more accurate to model testing costs separately from other development costs rather than at the total element line-item level.

Increasing the Granularity of Exploit Types

Similar to the more granular element estimates, it would be useful to enhance the robustness of exploit characterizations. The appendix provides additional detail on exploit types, but each has its nuances and could be a driver of cost.

Future model iterations could also consider how the exploit is implemented, whether it is dependent on other exploits, and the risks associated with a more complex acquisition.

Incorporating the Exploit Acquisition Type

The current framework does not make cost or schedule distinctions for how the cyber weapon was developed. For instance, some cyber weapon elements could be procured on the white or gray market and used as is; others could be procured and modified; and others could be completely developed in house. With additional historical data and low-level acquisition information, trends might show different cost and schedule estimates for the same element type, target environment, and complexity combination. An element developed completely in house could prove to be much more expensive and time-consuming than procuring something already on the market and modifying it to meet the same requirements. Currently, we do not have the data to make this claim, but as more data are collected, we might model this type of acquisition approach distinction.

Incorporating Additional Cost and Schedule Drivers

A potential modeling improvement might come from additional qualitative inputs. As understanding of the cyber weapon element acquisition space develops over time, we might learn of new cost and schedule drivers that we have not yet considered. Gaining a better understanding of what drives cost will allow us to improve the cost model.

Incorporating More-Robust ADL Decay Functions for Exploits

The ADL decay functions in the current model are largely notional and could be further developed to understand the factors that affect the ADL. Although the current Excel cost model allows for two ADLs, high or low, to account for broad differences between adversary types (e.g., a well-resourced nation-state versus a relatively unsophisticated terrorist organization), we have not conducted in-depth analysis on what additional factors might affect the ADL decay functions, what the appropriate exponent parameters are for the ADLs, and how many ADL functions might be appropriate to model and incorporate.

Incorporating Optimization into the Exploratory Simulation Model

The current exploratory model, as described in Chapter Three, can augment a baseline set of portfolios so that the user can determine how to achieve an operational objective or threshold requirement. A more sophisticated way to handle this would be to translate this exploratory model into an optimization. This optimization would be able to take an input set of capabilities (e.g., exploits), change their dates, duplicate them, and otherwise combine them to get a cost-effective, efficient frontier.[1] This would allow the user to determine the marginal cost of increasing the percentage of time in which the threshold or objective requirements are met. For example, it might allow the user to say, "The portfolio to hit the objective requirement for 80 percent of months costs $1 million, but the portfolio to hit the objective requirement for 90 percent of months costs $2 million." This sort of optimization would provide a more granular understanding of the best portfolio for a specific set of investments and thus the trade-offs between other parameters, but at the cost of significant additional computer time.

---

[1] In this scenario, the inputs are less like individual items in a portfolio and more like archetypes for various potential items that *could* be added to the portfolio. The optimization would then move them forward and backward in time and increase or decrease their quantity until the operational and/or cost goals are achieved.

Recommendations

Our findings point the following recommendations for JCW program leaders:

- Plan and budget for the significant uncertainty of the life span of vulnerabilities.
- Collect historical data (and plan to collect future data) on the cost of procuring and operationalizing exploits.

Over time, as more acquisition and execution data become available, we will be able to update the model and inputs to better predict future costs. We have identified seven ways to enhance the model: (1) incorporate historical data on cyber weapon element cost, (2) increase the granularity of cyber weapon element estimates, (3) increase the granularity of exploit types, (4) incorporate acquisition type for the exploit, (5) incorporate additional cost and schedule drivers, (6) incorporate more-robust ADL decay functions, and (7) incorporate optimization into the exploratory simulation model.

Type, Cost, and Life Span of Exploits

This appendix discusses the various characteristics of exploits in more detail.

Characteristics of Exploits

Remote Code Execution

As discussed in this report, exploits are most closely associated with the concept of access. System security architectures are designed to restrict the actions of unauthorized users, and exploits are the tools that bypass those architectures to provide additional privileges, permissions, and access to an attacker. An extreme example would be the ability to execute any arbitrary code or command (within the bounds of what is possible in the target computational environment) without restriction. Although exploits can enable a variety of actions, including access to data or details about system activity, file modification, and command execution, exploits that allow RCE are often the most valuable and sought after (MITRE, undated c). Exploits or chains of exploits that lead to RCE privileges can give an attacker more freedom to gain or expand access, install implants for persistence, and deploy malicious payloads.

Privilege Escalation

Even if an exploit allows RCE, that capability may be of little use if the code execution is confined to an isolated portion of a target system and cannot have a broader impact. Security architectures are also designed to limit an attacker's ability to traverse a network, modify files or processes, or execute code. Some exploits are designed to bypass these restrictions to obtain privileges, allowing an expanded set of permitted actions. An attacker could, for example, exploit software vulnerabilities to move from an unprivileged state to one involving user-level or root permissions (MITRE, 2021a). This is generally known as *privilege escalation*. It is often referred to as *local privilege escalation* because the attacker is traversing the local device or system.

Although the general category of privilege escalation encompasses a wide range of techniques, the set of strategies associated with avoiding or escaping isolated or containerized virtual environments is worth discussing in more detail. Whether primarily as a matter of functionality (e.g., virtual machines in a cloud environment) or security (e.g., browser applications operating in a sandbox), computing environments often isolate certain functions away from the host platform or OS. For a cyber weapon to eventually achieve its intended effect, the attacker will need to find some way to escape this isolation and obtain the capability for more-privileged actions (especially code execution) within the system hosting the isolated environment. Exploits that allow such an "escape to host" (often, more specifically, a virtual machine escape or sandbox escape) are often key privilege-escalation techniques required to achieve the follow-on objectives of a cyber weapon (MITRE, 2021b).

Tailoring Exploits to Targets and Other Weapon Elements

It is useful in the context of cyber weapon acquisition planning to discuss the number of exploits that may need to be acquired, but we must emphasize that exploits are not simply line-replaceable units that can be inserted into a cyber weapon. Unlike kinetic weapons, which can often be mass produced according to a design and applied equally well against a variety of physical targets, cyber weapons are often uniquely tailored to specific targets (Dykstra, Inglis, and Walcott, 2020). Each element of the weapon must be fitted to the target and to the other elements of the weapon if the weapon is to be effective. There may no value, for example, in acquiring a highly generalizable, highly reliable exploit if that exploit cannot help achieve the conditions and permissions needed to use implants or payloads designed for the target. Exploits therefore often need to be developed or acquired with specific implants and payloads in mind (Ablon and Bogart, 2017).

Exploits are often hard to generalize across platforms or network environments, independent of other characteristics of target organizations. Environments (e.g., Windows, mobile) are not equally valuable to an attacker and are not equally vulnerable to exploitation. Different configurations, software versions, network conditions, and even the skill of the operator can all affect the reliability of an exploit (Ablon and Bogart, 2017). Practices in target organizations can also have a significant impact on the expected reliability and effectiveness of exploits. Exploits are often distinguished by such descriptions as "one-click" or "zero-click," referring to the level of interaction they require from users in a target environment. If a target environment's organization regularly provides cybersecurity training to its users, this could lower the reliability of an exploit requiring user interaction (i.e., one-click). Moreover, as reflected in the cost framework's ADL, if an adversary has good organizational patch management practices, any exploitable zero-click vulnerabilities would be unlikely to persist for long, especially after a vulnerability becomes publicly known.

Determining the reliability of a single exploit might be much more important when a cyber weapon depends on many exploits working together in a chain to acquire the access and permissions needed to install an implant or deploy a payload. Individual exploits might only be as valuable as their dependent counterparts in a chain of actions that can deliver a cyber weapon's effects. In the same way that a rocket motor is not truly an element of a kinetic weapon until it is assembled as part of a missile or rocket, an exploit is a valuable component that is not truly useful until it is combined with the chained exploits, implants, and payloads needed to create a cyber weapon. That does not necessarily mean that "unchained" exploits are of no value. Indeed, it may be useful to have many components to choose from to better understand which components are missing and must be developed to assemble a tailored weapon that can deliver the desired effects. Having such a menu of exploits on the shelf can be useful, but these exploits could have a short shelf life (Ablon and Bogart, 2017).

Vulnerability Life Span

The varying types of exploits are dependent on vulnerabilities. We reviewed a variety of sources to characterize vulnerabilities and exploits to provide a better understanding of their life spans. Figure A.1 approximates key events in the life span. Vulnerabilities are often referred to as *in the wild* between their initial introduction and their identification by the owner of the software. The date of introduction can be difficult to ascertain because it could occur during a minor update, major update, earlier modification to the software, or some combination.

We identified several potential data sources to help us understand and characterize life spans. One of these sources, which is the basis for various research that has been performed on vulnerabilities and exploits, is MITRE's CVE program (MITRE, undated a). The CVE is an open-source list dating to 1999 containing identification numbers of more than 164,000 publicly known vulnerabilities, approximately 3,000 of which have been officially reviewed by the CVE program editorial board. Notably, the dataset does not include vulnerability introduction dates or detailed information beyond a brief description. The dates recorded in the dataset reflect the year that each vulnerability was disclosed. The CVE creates a standardized way of referencing vulnerabilities, facilitating a common link for additional vulnerability research efforts. The CVE list provides vulnerability identification, but independent research methods, such as those in Thomas et al., 2020, are required to ascertain more-detailed information on any given vulnerability.

The CVE also feeds into the National Vulnerability Database (NVD), run by the National Institute of Standards and Technology. The NVD analyzes CVEs by "aggregating data points from the description, references supplied and any supplemental data that can be found pub-

FIGURE A.1
Key Events in the Life Span of a Vulnerability

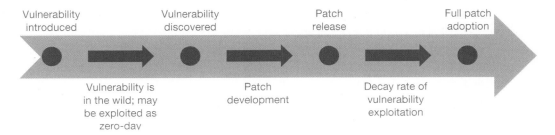

licly at the time" (National Institute of Standards and Technology, undated). However, this database also does not identify how long vulnerabilities were in the wild before disclosure.

Google Project Zero Analysis

Another source is Google Project Zero (Google, 2021). Project Zero data contain patch dates for 187 detected zero-day exploits since 2014. As a way of estimating how long an exploit was in the wild, we captured the length of time between the previous software update and the date of the patch. Using this method, we calculated operational times for 133 of the entries in the dataset. We then classified each entry into three broad target environment categories— desktop ($n = 112$), mobile ($n = 19$), and infrastructure ($n = 2$)—which we used within the exploratory model discussed in Chapter Three. We used two approaches to estimate the time an exploit was in the wild, which we refer to as *conservative* and *optimistic*. The conservative approach captured the time between the most recent software update and the patch date. The optimistic approach assumed that the vulnerability introduction date was the major version release preceding the update used in the conservative approach. Thus, if the conservative date coincides with a major version release, the optimistic date is the release date of the major version prior to that one. Each approach assumed that a software update was what gave rise to these vulnerabilities. To develop our estimates, we used publicly accessible product documentation to gather the following information:

- version format, typically given by a string of digits (e.g., XX.XXX.XX)
- major version characterization, typically given by the first set of digits in the version format (used to distinguish major from minor updates)
- update release history, which provides the date of release of all documented software updates for each product (also characterized as major or minor).

We define a *major update* as one associated with the release of a product's major version. For example, the Apple product iOS has the following version format: XX.X.X (e.g., 14.8.1). Major versions are characterized by the first two sets of digits; thus, the updates that introduced versions 14.7 and 14.8 are considered major updates. We established a similar definition for each product. Although there is no way to account for differences in version format between products, we kept the definitions of *major version* and *update* consistent with the Apple example, to the extent possible.

Table A.1 shows the estimated operational times for exploits categorized as either desktop or mobile using the optimistic approach described earlier. A caveat is that this dataset of publicly known zero-day exploits represents attack failures only, so average operational times would be longer if we could have included exploits that were still in the wild and undiscovered.

When incorporating these data into the exploratory model discussed in Chapter Three, we focused on the upper end of the operational dates to account for exploits out of this dataset that were still undiscovered. We used the mean times in Table A.1 as the mean values for the normal distributions representing the range of simulated operational times in cases with a

TABLE A.1

Optimistic Estimated Operational Time

Target Environment	Minimum Time (days)	Mean Time (days)	Maximum Time (days)
Desktop	9	139	650
Mobile	33	82	157

low ADL. We used the maximum times in the table as the means for the normal distributions in cases with a high ADL. We used the midpoint between those two values as the mean for normal distributions in cases with a medium ADL.

We also compared the range of estimated operational times against other attributes in the Project Zero data, including vendor name, product name, and exploit type. In making these comparisons, we included only the categories for which we could gather more than five data points. When exploring operational times by vulnerability type (see Figure A.2), we found Memory Corruption ($N = 99$) was the most observed and Use-After-Free ($N = 6$) the least observed, while Logic/Design Flaw had the largest interquartile range (IQR) (~258 days) and Use-After-Free the shortest (~21 days).

Grouping vulnerabilities by vendor (Figure A.3) showed that, collectively, Microsoft had the largest operational time range—almost twice as large as the second-largest range (637 days to Adobe's 339 days).

Grouping vulnerabilities by product (Figure A.4) further explains why Microsoft operational time ranges were so large. Chrome and Firefox, both of which are internet browsers, had the shortest operational time ranges, at 54 and 71 days, respectively. Internet Explorer (a Microsoft product) had the largest operational time range, at 630 days. The second-largest operational time belonged to Windows, another Microsoft product. According to documentation, Chrome and Firefox indeed received more frequent updates than the Microsoft products we observed.

Figure A.5 shows the estimated operational times for exploits targeting desktop and mobile products. In addition to total operational times, the Project Zero data can provide a look into the patch time between when a vulnerability is disclosed and when a patch is developed. We found that 75 entries contained a discover date and patch date, with the time between discovery and patch ranging from one to 182 days, with 24 days being the average.

Other Life Span Data Sources

We also reviewed several papers that analyzed various vulnerability and exploit data. These papers reported a range of exploit life spans. For example, Ablon and Bogart, 2017, determined that, on average, exploits had a life span of 6.9 years. Specifically, 25 percent had a life span of less than 1.5 years, and 25 percent endured for more than 9.5 years. Additionally, a 2020 analysis of 207 CVEs examined vulnerabilities targeting industrial control systems and found that temporal data were available for 165 out of the 207; these were in the wild

FIGURE A.2

Optimistic Estimated Operational Time, by Vulnerability Type

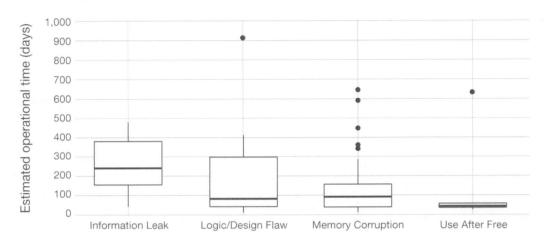

NOTES: Exploit types reflect original data categorization. The figure uses a black bar (mean value), a box (IQR), "whiskers" (indicating presence of data outside of IQR but within 1.5 × IQR), and black circles (outliers, defined as more than 1.5 × IQR).

FIGURE A.3

Optimistic Estimated Operational Time, by Vendor

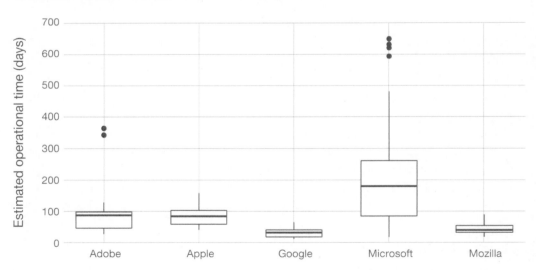

NOTE: Vendor types reflect original data categorization.

before disclosure for an average of 5.3 years and a range of 115–5,253 days (Thomas et al., 2020). That study also found little to no relationship between the life span of an exploit and its severity. Notably, because the CVE database does not include temporal data, as mentioned previously, the authors manually determined the length of time a vulnerability had existed by

FIGURE A.4

Optimistic Estimated Operational Time, by Product

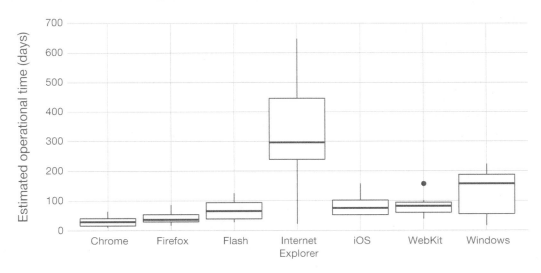

NOTE: Product types reflect original data categorization.

FIGURE A.5

Optimistic Estimated Operational Time, by Product Type

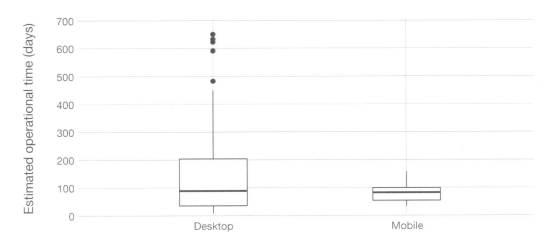

reviewing vendor websites to establish when affected versions and subsequent updates were published. Another 2017 analysis of 100,000 industrial control system devices found that 50 percent were patched within 60 days of vulnerability disclosure (Wang et al., 2017). A 2020 analysis of 60 vulnerabilities that were "either exploited or assigned a CVE number between Q1 2018 to Q3 2019" found that 58 percent were exploited as zero-day vulnerabilities; among those, researchers observed a range of zero to 1,153 days between the first exploit and the release of a patch (Metrick, Semrau, and Sadayappan, 2020). Finally, a 2020 analysis of 45,450 exploits found that "80% of public exploits are published before the CVEs are published" and that an exploit is published an average of 23 days before the vulnerability is reported to the CVE list (Chen, 2020).

Exploit Pricing

Exploit pricing was higher in 2019 than at any point previously, with a focus on exploits involving messaging apps, remote access, and on one-click or no-click exploits (Goodin, 2019). The price of an exploit can generally be linked to how long the exploit has existed and whether a patch for the vulnerability has been released (Trend Micro, 2021). However, several factors can influence exploit cost (Ruef, 2016):

- target popularity (sometimes detrimental)
- exploit exclusivity
- exploit quality
- attack reliability
- penetration ability
- attack options.

Other factors have an inverse relationship with exploit price:

- vulnerability existence disclosure
- vulnerability detail disclosure
- alternative exploit development
- countermeasures released.

Overall, we found that vulnerability price data are limited, with Zerodium offering the most comprehensive dataset. Zerodium, established in 2015, is a company that acquires zero-day exploits and brokers them to government organizations (Zerodium, undated). To obtain data, we used the Wayback Machine to view versions of the webpage for each year and built a dataset of exploit costs for the years 2015–2019 (Internet Archive, undated). Based on the data, we identified the following:

- **average price floor:** the lowest price, on average, for that type of exploit, in thousands of dollars
- **average price ceiling:** the highest price, on average, for that type of exploit, in thousands of dollars
- **average price:** the average price overall for that type of exploit, in thousands of dollars.

We found that certain exploits command a high price, particularly exploits of mobile platforms, messaging apps, and zero-click and remote exploits. Notably, although iOS exploits were historically more valuable, Android exploits have recently overtaken iOS for top bounty. Specifically, Zerodium initially offered $0.5 million and $0.1 million for iOS and Android exploits, respectively, in 2015 but offered $2.5 million for Android RCE in September 2019 (Greenburg, 2019).

Abbreviations

ADL	adversary defense level
CDF	cumulative distribution function
CVE	Common Vulnerability and Exposures
DoD	U.S. Department of Defense
FY	fiscal year
ID	identification
IQR	interquartile range
JCW	Joint Cyber Weapons
MoM	method of moments
O&M	operations and maintenance
OS	operating system
RCE	remote code execution
SME	subject-matter expert
SWP	software acquisition pathway
TY	then year

References

Ablon, Lillian, and Andy Bogart, *Zero Days, Thousands of Nights: The Life and Times of Zero-Day Vulnerabilities and Their Exploits*, Santa Monica, Calif.: RAND Corporation, RR-1751-RC, 2017. As of November 22, 2021:
https://www.rand.org/pubs/research_reports/RR1751.html

Arena, Mark V., Robert S. Leonard, Sheila E. Murray, and Obaid Younossi, *Historical Cost Growth of Completed Weapon System Programs*, Santa Monica, Calif.: RAND Corporation, TR-343-AF, 2006. As of November 22, 2021:
https://www.rand.org/pubs/technical_reports/TR343.html

Ballard, Barclay, "Only a Tiny Percentage of Security Vulnerabilities Are Actually Exploited in the Wild," webpage, TechRadar, February 19, 2021. As of November 22, 2021:
https://www.techradar.com/news/only-a-tiny-percentage-of-security-vulnerabilities-are-actually-exploited-in-the-wild

Bellovin, Steven M., Susan Landau, and Herbert S. Lin, "Limiting the Undesired Impact of Cyber Weapons: Technical Requirements and Policy Implications," *Journal of Cybersecurity*, Vol. 3, No. 1, March 2017, pp. 59–68.

Bevington, Philip R., *Data Reduction and Error Analysis for the Physical Sciences*, McGraw-Hill, 1969.

Chen, Jay, "The State of Exploit Development: 80% of Exploits Publish Faster Than CVEs," blog post, Palo Alto Networks, August 26, 2020. As of November 22, 2021:
https://unit42.paloaltonetworks.com/state-of-exploit-development/

Department of Defense Instruction 5000.02, *Operation of the Adaptive Acquisition Framework*, Washington, D.C., January 23, 2020.

Department of Defense Instruction 5000.87, *Operation of the Software Acquisition Pathway*, Washington, D.C., October 2, 2020.

Dykstra, Josiah, Chris Inglis, and Thomas S. Walcott, "Differentiating Kinetic and Cyber Weapons to Improve Integrated Combat," *Joint Force Quarterly*, No. 99, 4th Quarter 2020, pp. 116–123.

Goodin, Dan, "Zeroday Exploit Prices Are Higher Than Ever, Especially for iOS and Messaging Apps," webpage, Ars Technica, January 7, 2019.

Google, "0day 'In the Wild,'" spreadsheet, updated May 5, 2021. As of November 22, 2021:
https://docs.google.com/spreadsheets/d/1lkNJ0uQwbeC1ZTRrxdtuPLCIl7mlUreoKfSIgajnSyY/view#gid=0

Greenburg, Andy, "Why 'Zero Day' Android Hacking Now Costs More Than iOS Attacks," *Wired*, September 3, 2019. As of November 22, 2021:
https://www.wired.com/story/android-zero-day-more-than-ios-zerodium

Internet Archive, "Wayback Machine," webpage, undated. As of November 22, 2021:
https://archive.org/web

Libicki, Martin C., Lillian Ablon, and Tim Webb, *The Defender's Dilemma: Charting a Course Toward Cybersecurity*, Santa Monica, Calif.: RAND Corporation, RR-1024-JNI, 2015. As of August 7, 2022:
https://www.rand.org/pubs/research_reports/RR1024.html

Lockheed Martin, *Gaining the Advantage: Applying Cyber Kill Chain Methodology to Network Defense*, Bethesda, Md., 2015. As of November 4, 2021:
https://www.lockheedmartin.com/content/dam/lockheed-martin/rms/documents/cyber/Gaining_the_Advantage_Cyber_Kill_Chain.pdf

Metrick, Kathleen, Jared Semrau, and Shambavi Sadayappan, "Think Fast: Time Between Disclosure, Patch Release and Vulnerability Exploitation—Intelligence for Vulnerability Management, Part Two," blog post, *FireEye Mandiant Threat Intelligence,* April 13, 2020. As of November 22, 2021:
https://www.fireeye.com/blog/threat-research/2020/04/time-between-disclosure-patch-release-and-vulnerability-exploitation.html

Microsoft, "HAFNIUM Targeting Exchange Servers with 0-Day Exploits," webpage, March 2, 2021a. As of November 22, 2021:
https://www.microsoft.com/security/blog/2021/03/02/hafnium-targeting-exchange-servers

Microsoft, "Analyzing Attacks Taking Advantage of the Exchange Server Vulnerabilities," webpage, March 25, 2021b. As of November 22, 2021:
https://www.microsoft.com/security/blog/2021/03/25/analyzing-attacks-taking-advantage-of-the-exchange-server-vulnerabilities

MITRE Corporation, "Common Vulnerabilities and Exposures Program," website, undated a. As of November 22, 2021:
https://cve.mitre.org/

MITRE Corporation, "Common Vulnerabilities and Exposures Program: Glossary," webpage, undated b. As of November 22, 2021:
https://www.cve.org/ResourcesSupport/Glossary

MITRE Corporation, "MITRE ATT&CK," website, undated c. As of November 22, 2021:
https://attack.mitre.org

MITRE Corporation, "Exploitation for Privilege Escalation," webpage, last updated April 22, 2021a. As of November 22, 2021:
https://attack.mitre.org/techniques/T1068

MITRE Corporation, "Escape to Host," webpage, last updated October 15, 2021b. As of November 22, 2021:
https://attack.mitre.org/techniques/T1611

National Initiative for Cybersecurity Careers and Studies, "Glossary—Explore Terms: A Glossary of Common Cybersecurity Words and Phrases," webpage, June 9, 2022. As of June 30, 2022:
https://niccs.cisa.gov/about-niccs/cybersecurity-glossary

National Institute of Standards and Technology, "National Vulnerability Database: General Information," webpage, undated. As of November 22, 2021:
https://nvd.nist.gov/general

Naval Center for Cost Analysis, *Joint Agency Cost Schedule Risk and Uncertainty Handbook*, 2014.

Naval Center for Cost Analysis, "Joint Inflation Calculator," webpage, March 2021. As of April 1, 2021:
https://www.ncca.navy.mil/tools/inflation.cfm

Office of the General Counsel, *Department of Defense Law of War Manual*, Washington, D.C.: U.S. Department of Defense, December 2016. As of June 30, 2022:
https://ogc.osd.mil/Portals/99/law_war_manual_december_16.pdf

Pasagian, Arthur J., "Decision Memorandum for Joint Cyber Weapons: Payloads Exploits, and Implants Program Use of the Software Acquisition Pathway and Entry into the Planning Phase," memorandum, October 22, 2021.

Public Law 116-92, National Defense Authorization Act for Fiscal Year 2020, December 20, 2019.

The R Project for Statistical Computing, homepage, undated. As of November 22, 2021:
https://www.r-project.org

Ruef, Marc, "Exploit Pricing—Analysis of the Market in Digital Weapons," *SCIP* blog, 2016. As of November 22, 2021:
https://www.scip.ch/en/?labs.20161013

StatCounter, "Browser Market Share Worldwide, December–January 2019," dataset, undated. As of November 22, 2021:
https://gs.statcounter.com/browser-market-share/all/worldwide/2019

Stockton, Paul N., and Michele Golabek-Goldman, "Curbing the Market for Cyber Weapons," *Yale Law and Policy Review*, Vol. 32, No. 1, 2013, pp. 239–266.

Thomas, Richard J., Joseph Gardiner, Tom Chothia, Emmanouil Samanis, Joshua Perrett, and Awais Rashid, "Catch Me if You Can: An In-Depth Study of CVE Discovery Time and Inconsistencies for Managing Risks in Critical Infrastructures," *Proceedings of the 2020 Joint Workshop on CPS&IoT Security and Privacy*, New York: Association for Computing Machinery, November 2020, pp. 49–60.

Trend Micro, "Trends and Shifts in the Underground N-Day Exploit Market," webpage, July 13, 2021. As of November 22, 2021:
https://www.trendmicro.com/vinfo/us/security/news/vulnerabilities-and-exploits/trends-and-shifts-in-the-underground-n-day-exploit-market

Wang, Brandon, Xiaoye Li, Leandro P. Aguiar, Daniel S. Menasche, and Zubir Shafiq, "Characterizing and Modeling Patching Practices of Industrial Control Systems," *Proceedings of the ACM on Measurement and Analysis of Computing Systems*, Vol. 1, No. 1, June 2017, article 18.

Zerodium, "Zerodium Exploit Acquisition Program," webpage, undated. As of November 22, 2021:
https://zerodium.com/program.html